T0135103

Electron Microscopy and Analysis

Electron Microscopy and Analysis

Third edition

Peter J. Goodhew
University of Liverpool, UK

John Humphreys
Manchester Materials Science Centre, UK

Richard Beanland
Marconi Materials Technology, Towcester, UK

London and New York

Published 2001 by
Taylor & Francis
11 New Fetter Lane, London EC4P 4EE

Simultaneously published in the USA and Canada
by Taylor & Francis Inc
29 West 35th Street, New York, NY 10001

Taylor & Francis is an imprint of the Taylor & Francis Group

© 2001 Peter J. Goodhew, John Humphreys and Richard Beanland
First edition published 1975 by Wykeham Publications (London) Ltd
Second edition published 1988 by Taylor & Francis

British Library Cataloguing in Publication Data
A catalogue record for this book is available
from the British Library

Library of Congress Cataloging in Publication Data
Goodhew, Peter J.
    Electron microscopy and analysis / Peter J. Goodhew, John Humphreys, Richard Beanland – 3rd ed.
        p. cm
    Includes bibliographical references (p.).
    1. Electron microscopy.   I. Humphreys, John   II. Beanland, Richard.   III. Title.

QH212.E4 G62 2000
502'.8'25–dc21   00-037716

ISBN   0 7484 0968 8

Contents

Acronyms

AEM	Analytical electron microscopy
AES	Auger electron spectroscopy
AFM	Atomic force microscopy
ALCHEMI	Atom location by channelling enhanced microanalysis
APFIM	Atom probe field ion microscopy
CBED	Convergent beam electron diffraction
CBIM	Convergent beam imaging
CTEM	Conventional TEM
EBSD	Electron back scattering diffraction
ECP	Electron channelling pattern
EDX (or EDS)	Energy dispersive X-ray analysis
EPMA	Electron probe microanalysis
ESCA	Electron spectroscopy for chemical analysis ($=$XPS)
FEG	Field emission gun
FIB	Focused ion beam
FIM	Field ion microscopy
HOLZ	High order Laue zone
HREM	High resolution electron microscopy
HVEM	High voltage electron microscopy
LACBED	Large angle convergent beam diffraction
LEED	Low energy electron diffraction
LEEM	Low energy electron microscopy
LIMA	Laser ionization mass analysis
MFM	Magnetic force microscopy
MS	Mass spectrometry
PEELS	Parallel electron energy loss spectrometry
POSAP	Position sensitive atom probe
RBS	Rutherford backscattering
REM	Reflection electron microscopy
SACP	Selected area channelling pattern
SAD	Selected area diffraction
SAM	Scanning Auger microscopy/Scanning acoustic microscopy

SEM	Scanning electron microscopy
SIMS	Secondary ion mass spectrometry
SPM	Scanning probe microscopy
STEM	Scanning transmission electron microscopy
STM	Scanning tunnelling microscopy
TEM	Transmission electron microscopy
TOFSIMS	Time of flight SIMS
UPS	Ultraviolet photoelectron spectroscopy
WEDX	Windowless EDX
WDX	Wavelength dispersive X-ray spectrometry
XPS	X-ray photoelectron spectrometry (= ESCA)
ZAF	Atomic number, absorption and fluorescence correction

Preface

It has been gratifying to discover in how many countries of the world the first two editions of this slim textbook have been read. It is now twelve years since the second edition was written in 1987 and this has been a period of rapid development in both electron microscopy itself and in related techniques. Many of the most visible developments are associated with the introduction of computer control into microscopy – it is now commonplace to find an SEM driven from a keyboard or mouse rather than from dedicated knobs on the instrument fascia. Other developments have only become feasible since the advent of cheap high power computing; confocal light microscopy and scanning probe microscopy are both good examples.

In writing this third edition we have been conscious that the basic principles of SEM and TEM have not changed, whereas the outward appearance of many microscopes has. During the 1980s scanned probe microscopies have also grown up. They are no longer specialized laboratory instruments but some of them (e.g. AFM) have become off-the-shelf everyday imaging tools. We have tried to reflect these changes without removing the material which leads to fundamental understanding of microscopy itself.

The book now covers a wide range of topics and it is a pleasure to acknowledge the help of many colleagues who offered help in particular areas, especially Dr Helen Davock who provided several of the illustrations and Anne Beckerlegge who imposed order on our chaos.

Peter J. Goodhew, John Humphreys and Richard Beanland
Liverpool, Manchester and Towcester

Chapter 1

Microscopy with light and electrons

1.1 Introduction

A microscope is an optical system which transforms an 'object' into an 'image'. We are usually interested in making the image much larger than the object, that is magnifying it, and there are many ways in which this can be done. This book deals with several sophisticated techniques for magnifying images of very small objects by large amounts, but many of the principles involved are just the same as those which have been developed for light microscopes over the past 400 years. The concepts of resolution, magnification, depth of field and lens aberration are very important in electron microscopy and so we deal with them in this first chapter in the more familiar context of the light microscope. When we consider electron microscopes in later chapters it will be found that instead of becoming more complicated, many areas of the subject become simpler because we are dealing with electrons rather than light. Thus although apparently more complex, and certainly much more expensive, electron microscopes are almost as easy to understand (in principle) as their humble stable-mate, the magnifying glass.

The techniques which are given the most detailed coverage in this book are scanning electron microscopy (SEM), transmission electron microscopy (TEM) and the analytical techniques which are made available by using them. At the simplest level an SEM can be thought of as providing images of external morphology, rather similar in appearance to those formed by the eye, while a TEM probes the internal structure of solids and gives us access to micro-structural or ultrastructural detail not familiar to the human eye. In both cases several different types of image can be formed. Consequently it is necessary to understand not only how such microscopes work, but also how to interpret the images which they produce. This is particularly true for the TEM, and accounts for the length of Chapter 4.

It will become apparent that there is much more to both SEM and TEM than is implied by the preceding paragraph, since there is almost infinite scope for control of the imaging processes to reveal specific types of detail in a specimen. In addition, the potential for analysis of local regions of a specimen while it is being examined in an electron microscope has been exploited very widely since the first edition of this book was published. Chapter 6 deals in some detail with several of the analytical methods such as X-ray and electron spectrometry, which are now standard facilities on a modern microscope. It is also easy to obtain structural as well as chemical information by using electron diffraction in either an SEM or a TEM and this topic is treated in Chapter 3. Altogether the modern family of electron microscopes are extremely versatile tools for revealing the nature and behaviour of matter.

Several of the concepts which are essential to the understanding of electron microscopy are common to any imaging system, and many of these ideas will have first been met in the context of the light microscope. In the remainder of this chapter we therefore deal with some general features of imaging systems and then introduce the ideas of magnification, resolution and lens aberrations as they apply to simple and familiar light-optical systems.

1.2 Methods of image formation

There are three basic ways in which an image can be formed. Perhaps the simplest to imagine is the *projection* image, of which the commonest example is the formation of shadows when an object is placed in front of a point source of illumination, as shown in Figure 1.1. The second type of image is formed by conventional lens systems, as for example in Figure 1.2, and we shall call this an *optical* image. This is not a strictly accurate term, since optical simply means 'involving light' and one of the key lessons of this book is that similar images can be formed using electrons or ions. We will try to use the terms 'electron-optical' and 'ion-optical' where they are appropriate.

Both projection and optical images are formed in parallel, that is all parts of the image are formed essentially simultaneously. However the third type of image we need to consider is the *scanning* image, in which each point of the picture is presented serially. The best-known example of this type of image is a television picture, in which several thousand picture points are displayed con- secutively, but the process is repeated with such a high frequency that the image appears to the eye in its entirety.

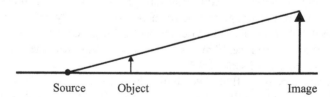

Source   Object   Image

Figure 1.1 The formation of a projection (or shadow) image. Each point in the object is projected directly at the equivalent point in the image.

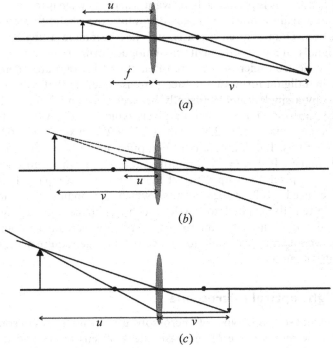

Figure 1.2 Ray diagrams illustrating the formation of an image by a single lens of focal length f. An animated version of this ray diagram can be found in MATTER: Introduction to Electron Microscopes.

In later chapters we will devote a great deal of attention to one optical technique, the TEM, and to one scanning technique, the SEM. We will also, for comparison, describe a projection technique, the Field Ion Microscope (FIM).

1.3 Pixels

One of the most important ideas concerned with images arises from the scanning image as typified by television. A European TV image contains about 700×625 discrete points and it is fairly obvious that the smallest piece of information about the image is contained in one of these *picture points*. They are generally called *pixels*, which is short for picture element. A single domestic TV picture therefore consists of more than 200 000 pixels, each of which can be of a different intensity or colour. The smallest detail which can possibly be shown in the image is a single pixel in size, that is 1/625 of the screen height for a TV image. The idea of the pixel arose from consideration of scanned images but it turns out to be universally applicable to images however they are formed. This is particularly relevant when an image is to be stored by a computer, and again it must be broken down into the smallest necessary units of information. Domestic television is only just becoming a 'digital' medium and the signals are currently sent in analogue form. However the images produced by electron microscopes are often stored in computer memory and need to be in a digital form, that is each pixel is coded so that its brightness is represented by a single number (usually between zero and 255). Such images are often composed of a number of pixels which is a power of two, and common image sizes are 256×256 ($= 2^8 \times 2^8$) pixels or 1024×1024 ($= 2^{10} \times 2^{10}$) pixels. Large amounts of computer memory are then needed to store such images. If 256 ($= 2^8$) brightness levels (known as grey levels) are permitted, each pixel takes up 8 bits of memory and a complete 1024×1024 pixel image needs $1024 \times 1024 \times 8$ bits, which for many computers is 1 megabyte (often abbreviated to 1 MB). Such an image will just fit on the conventional floppy disc of a microcomputer. However, image compression techniques are making it possible to reduce the storage requirement, often by an order of magnitude.

1.4 The light-optical microscope

Both optical and scanning types of microscope usually use lenses in some form, so we will now review some of the basic ideas of lens optics and define the necessary terminology. The simplest optical microscope, which has been in use since the early seventeenth century, is a single convex lens or 'magnifying glass'. The ray diagram for this is shown in Figure 1.2 and serves to illustrate the concepts of focal length, f, and magnification, M. The image is magnified, real and inverted if the object distance u (between lens and object) is between f and $2f$, as shown in Figure 1.2(a) and in MATTER: Introduction to Electron Microscopes. The image is erect but virtual if the object is within the focal distance (i.e. the object distance is between zero and f, Figure 1.2(b)). If an image is to be recorded on a photographic plate or viewed on a screen then it must be real, and therefore we will not be concerned with optical arrangements which give rise to virtual final images.

If the object is further from the lens than $2f$ (Figure 1.2(c)) the image is de-magnified, that is the magnification is less than unity. Notice that Figures 1.2(a) and 1.2(c) are essentially the same (although drawn backwards) if we interchange the words 'object' and 'image'. This illustrates one of the important features of an optical system – its effect on the light rays does not depend on the direction in which they are supposed to be travelling. This 'principle of reciprocity' was propounded by Helmholtz in 1886 and will be of use when we consider the scanning transmission electron microscope (STEM) in Chapter 7.

The conclusions drawn above about the behaviour of a convex lens of focal distance f are summarized in the thin lens equation

$$\frac{1}{f} = \frac{1}{u} + \frac{1}{v} \qquad (1.1)$$

where u is the 'object distance' (the distance from the lens to the object) and v is the 'image distance'. Figure 1.2(b) shows that, by similar triangles, the magnification M produced by the single lens is given by v/u. Substitution in the lens equation gives

$$M = \frac{f}{u-f} = \frac{v-f}{f} \qquad (1.2)$$

from which it can be deduced that for large magnification $u - f$ must be small and positive. This is achieved by placing the object just outside the focal point of the lens.

Magnification of an object without severe distortion is very limited using a single lens. Strictly the image in Figure 1.2(a) should be curved so that all points on it are equidistant from the lens centre. If the magnification is high this effect is considerable and the image seen in any one plane will appear distorted. For high magnifications therefore, combinations of lenses are used so that the total magnification is achieved in two or more stages. A simple two-stage photomicroscope will have the ray diagram shown in Figure 1.3.

The first lens, the *objective*, provides an inverted image at B with a magnification $(v_1 - f_1)/f_1$ and the second lens, the *projector*, gives a final upright image at a further magnification of $(v_2 - f_2)/f_2$. The image is viewed on a screen or recorded on a photographic plate at C with a total magnification of

$$M = \frac{(v_1 - f_1)(v_2 - f_2)}{f_1 f_2} \qquad (1.3)$$

If higher magnifications are required it is quite straightforward to add a second projector lens to provide a third stage of magnification.

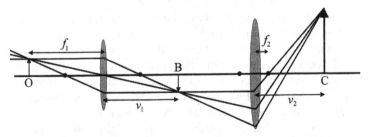

Figure 1.3 The ray diagram of a simple two-stage projection microscrope. The object is at O and the final image at C, with an intermediate image at B.

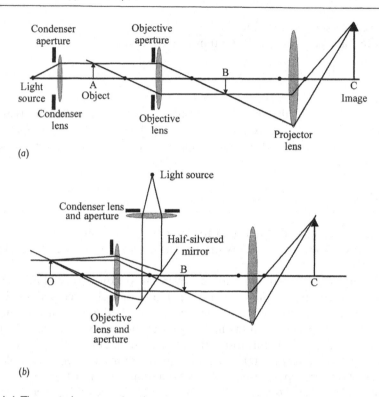

Figure 1.4 The optical systems for the two common types of projection microscope. (a) Transmission illumination, (b) Reflected illumination.

We have so far assumed that the object itself is self-luminous, and we have therefore shown the 'rays' starting at the object and ending at the viewing screen. In practice we are rarely able to look at this sort of specimen and we must illuminate it with light from a convenient source. We are now forced to consider whether the object is mainly transparent, in which case we illuminate it from behind, or whether it is opaque, in which case we must illuminate it from the front. Thus, immediately, we have a division into two classes of optical microscope: the biologist who needs to look at very thin sections of tissue uses a transmission arrangement such as that shown in Figure 1.4(a), while the materials scientist or geologist who needs to examine the surface structure of a solid specimen uses a reflection arrangement as shown in Figure 1.4(b). The same two types of electron-optical arrangement arise in electron microscopy, leading to TEM and SEM instruments. In this case both types of instrument are used in almost all fields of science.

The essential parts of any illumination system are a light source and a condenser system. The condenser is necessary to collect the light which is diverging from the source and to direct it at the small area of the specimen

Figure 1.5 An optical micrograph of a polished specimen of Al–16% Si. (Monica Hughes, University of Liverpool)

which is to be examined. This serves two purposes; it makes the object appear brighter so that it can be seen more easily (also improving its *contrast*) and it also enables the microscopist to control the angle at which the illumination arrives at the specimen. The beam can be made to converge on the specimen or can illuminate it with parallel rays. It will be shown later that in electron microscopy the concepts of contrast and convergence angle are rather important and we will deal with them in greater detail in Chapters 3, 4 and 5.

In early light microscopes the sun or ordinary diffuse daylight was used as a source and a concave mirror was used to direct the light towards the specimen. For many purposes this is adequate but for more demanding work it is more usual to find a built-in light source and a condenser lens as shown in Figure 1.4. With the addition of two variable apertures near the condenser lens and the objective lens it is possible to control the area of specimen which is illuminated and the angular spread of the light collected from the specimen. With a well-made microscope, micrographs such as that shown in Figure 1.5 can be taken.

1.5 Magnification

In principle it is possible to make a light microscope which will produce any selected magnification. However, since for convenience the instrument should be compact, without too many adjustments, it is usual to alter f_1 or f_2 in equation 1.3 rather than v_1 or v_2. This means that in order to change the magnification, one lens is usually exchanged for another with a different focal length, giving a limited set of fixed magnifications. The alternative is to alter the distances between all the components of the microscope and this is generally less convenient. We will see later that this problem does not arise in electron microscopes, where all the parameters are more easily adjusted.

Although it was stated in the last section that the total magnification of the microscope can easily be increased by adding additional lenses it turns out that for the vast majority of purposes the two-lens system shown in Figure 1.3 is quite sufficient. The reason for this is simple; the smallest details which can usefully be distinguished in a light microscope are about 200 nm in size $(2 \times 10^{-7} \text{m}: 1000 \text{nm} = 1 \text{μm}; 1000 \text{μm} = 1 \text{mm})$. The reason for this limit is discussed in the next section but for the moment let us consider its implication. The unaided human eye can easily detect detail only 0·2 mm in size. Therefore there is very little point in magnifying the smallest details which can be resolved (200 nm) up to a larger size than 0·2 mm (200 μm). Thus any magnification greater than 1000× only makes the details bigger. We cannot make finer details visible by magnifying the image an extra ten times. An example of this 'empty magnification', as it is called, is shown in Figure 1.6. The first micrograph has a magnification of 70× and we see a lot of detail. Magnifying this several times more, to 300× reveals more detail. However a further stage of magnification to 1400× or higher shows us no more; the features are further apart but no clearer. If a large display is needed, for example in order to view the micrograph at a distance, it is more sensible to enlarge a 1000× micrograph photographically than to build a microscope capable of higher magnifications. Now it is relatively easy to provide magnifications of 1000× with only the two-lens system of Figure 1.3, for example using an 80× objective lens and a 15× projector lens. Consequently it is not necessary to build a light microscope with three or more stages of magnification, since this will not improve the resolution but will rather degrade it by introducing extra aberrations (see section 1.8). However the scanning electron microscope has inherently better resolution and it makes sense to use it at higher magnifications, as Figure 1.6 shows.

1.6 Resolution

In order to compare the electron microscope with the light microscope we need to know what factors control the resolution (often called resolving power) which we will define as the closest spacing of two points which can clearly be seen through the microscope to be separate entities. Notice that this is not necessarily the same as the smallest point which can be seen with the microscope, which will often be smaller than the resolution limit.

Even if all the lenses of the microscope were perfect and introduced no distortions into the image, the resolution would nevertheless be limited by a diffraction effect. Inevitably in any microscope the light must pass through a

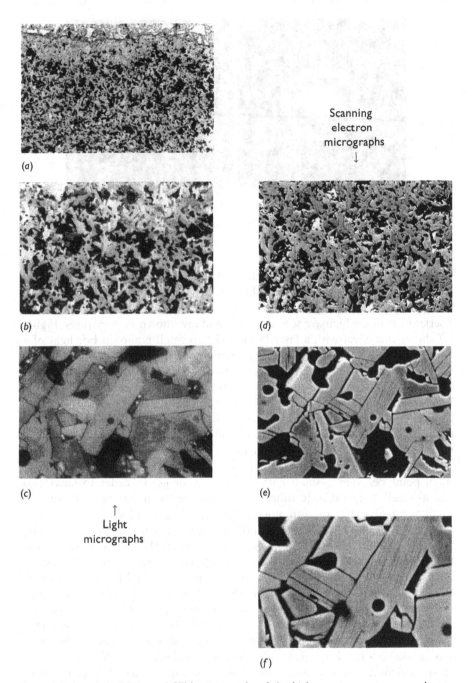

Scanning
electron
micrographs
↓

(a)

(b)

(d)

(c)

(e)

↑
Light
micrographs

(f)

Figure 1.6 A series of light and SEM micrographs of the high temperature superconductor barium yttrium copper oxide at increasing magnification. Original magnifications (a) 70 ×; (b) and (d) 300 ×; (c) and (e) 1400 × and (f) 2800 ×.

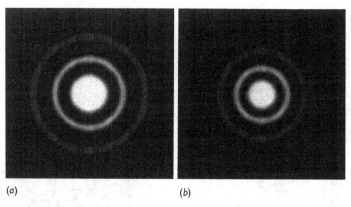

(a)   (b)

Figure 1.7 Airy rings resulting from the diffraction of a laser beam by small pinholes. (a) 75 μm diameter; (b) 100 μm diameter.

series of restricted openings – the lenses themselves or the apertures shown in Figure 1.4. Wherever light passes through an aperture, diffraction occurs so that a parallel beam of light (which would be seen as a spot) is transformed into a series of cones, which are seen as circles and are known as Airy rings. Figure 1.7 shows this effect with a laser beam and two small pinholes. For light of a given wavelength the diameter of the central spot is inversely proportional to the diameter of the aperture from which the diffraction is occurring. Consequently the smaller the aperture, the larger is the central spot of the Airy disc. We have used very small apertures in order to make the Airy disc clearly visible but the same effects occur from the relatively larger apertures found in light microscopes. The diffraction effect limits the resolution of a microscope because the light from every small point in the object suffers diffraction, particularly by the objective aperture, and even an infinitely small point becomes a small Airy disc in the image. In order to make this disc as small as possible, in other words to make the image of each point as small as possible, the aperture must be as large as is feasible.

Now let us consider the resolution of the microscope in more detail, starting with the Airy disc. Figure 1.8 shows the variation of the light intensity across the series of rings which make up the disc. The central spot is very much more intense than any other ring and in fact contains 84% of all the light intensity. Consequently for many purposes the rings can be ignored and we can assume that all the light falls in a spot of diameter d_1, where $d_1 \propto 1/$(aperture diameter). Consider how far apart two of these spots must be in the image before they are distinguishable as two – this distance is the resolution which was defined earlier. Lord Rayleigh proposed a criterion which works well in most cases and has been used extensively ever since; when the maximum of intensity of an Airy disc coincides with the first minimum of the second, then the two points can just be distinguished. This is illustrated in Figure 1.9, from

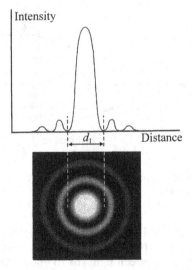

Figure 1.8 The variation of light intensity across a set of Airy rings. Most of the intensity (84%) lies within the first ring, that is within a spot of diameter d_1.

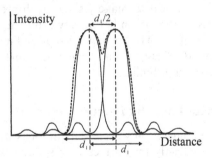

Figure 1.9 The intensity of the Airy rings from two neighbouring pinholes. The intensity distributions from each of the pinholes separately are shown as solid lines; the combined profile from the two pinholes acting together is shown dotted. At the Rayleigh resolution limit, as shown here, the maximum intensity from one pinhole coincides with the first minimum from the other. This gives a resolution limit of $d_1/2$.

which it can be seen that the resolution limit is $d_1/2$. Microscope apertures are normally referred to in terms of the semi-angle, α, which they subtend at the specimen (Figure 1.10). It is then possible to derive from diffraction theory (see any text on optics) the relationship;

$$r_1 = \frac{d_1}{2} = \frac{0.61\lambda}{\mu \sin \alpha} \tag{1.4}$$

where λ is the wavelength of the light and μ is the refractive index of the medium between the object and the objective lens. The product, $\mu \sin \alpha$ is usually called the *numerical aperture* (NA).

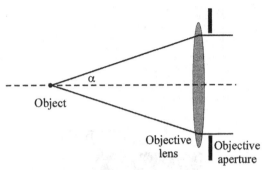

Figure 1.10 The definition of the half-angle, α, subtended by an aperture (in this case the objective aperture).

In order to obtain the best resolution (i.e. the smallest r_1) it is obviously possible to decrease λ or increase μ or α. With a light microscope λ can be decreased to 400 nm by using green light (or to about 200 nm if it is possible to use ultraviolet light); sin α can be increased towards 1 by using as large an aperture as possible and μ can be increased by using an oil immersion objective lens. However it is impractical to make μ sin α much greater than about 1·6 since sin α must be less than unity and even very exotic materials are limited to a refractive index of about 1·7. The absolute resolution limit using green light is therefore about 150 nm (0.15 μm). Even sophisticated image processing cannot improve on this fundamental limit.

1.7 Depth of field and depth of focus

In any microscope the image is only accurately in focus when the object lies in the appropriate plane (strictly the surface of a sphere). If part of the object being viewed lies above or below this plane then the equivalent part of the image will be out of focus. The range of positions for the object for which our eye can detect no change in the sharpness of the image is known as the *depth of field*. In most microscopes this distance is rather small and therefore in order to produce sharp images the object must be very flat. If a non-flat object (or a transparent object of appreciable thickness) is viewed at high magnification using a light microscope then some out-of-focus regions will be seen. This is a useful feature if we wish to accentuate certain parts of the image at the expense of others but is a grave disadvantage if we would like to see all parts of a three-dimensional object clearly.

In the 1990s clever optical design and the use of scanning have led to the development of confocal light microscopes which exploit the intrinsic narrow depth of field to build up a 'three-dimensional' image which when viewed is in focus over a range of depths. This is described in more detail in Chapter 7.

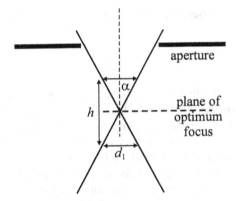

Figure 1.11 The depth of focus of an optical system, *h*, is the distance from the plane of optimum focus within which the beam diverges by no more than the spot diameter d_1. d_1 will be limited by diffraction and aberrations.

The depth of field can be estimated from Figure 1.11, which shows rays converging at the specimen. Since the diffraction effect described in section 1.6 will limit the resolution at the specimen to r_1 (given by equation 1.4) it will not make any difference to the sharpness of the image if the object is anywhere within the range *h* shown in Figure 1.11. Simple geometry then gives:

$$h = \frac{0.61\lambda}{\mu \sin \alpha \tan \alpha} \tag{1.5}$$

from which it is evident that the only effective way to increase the depth of field is to decrease the convergence angle, which is controlled in most cases by the objective aperture, as Figure 1.10 shows. Notice that conditions which maximize the depth of field simultaneously make the resolution worse (equation 1.4). For a light microscope, where α might be in the region of 45 degrees, the depth of field is not very different from the resolution. Even if the objective convergence is limited to 5 degrees the depth of field will only be about 40 μm, while the resolution will then be limited to about 3 μm.

It will become apparent later that the use of electrons for microscopy brings a number of advantages, among which are an improvement in *both* resolution and depth of field. The reason for this is that high energy electrons have a much smaller wavelength than light (see Chapter 2) and the microscopes are usually operated with very small values of α.

A term which is often confused with depth of field is the *depth of focus*. This refers to the range of positions at which the image can be viewed without appearing out of focus, for a fixed position of the object. Depth of focus is often not as important as the depth of field to a microscopist, but in any case tends to be larger, as the following calculation shows.

Equation 1.1 can be differentiated (for constant focal length) to give:

$$\frac{dv}{du} = \frac{-v^2}{u^2} = -M^2 \qquad (1.6)$$

This shows that dv, the effective shift in image position, is related to du, the change in position of the object, via the square of the magnification. The negative sign arises because the shifts are in opposite directions, which does not concern us. Thus if du is set to be the depth of field, calculated perhaps from equation 1.5, the equivalent depth of focus is a factor M^2 bigger. At any reasonable magnification the depth of focus will therefore be large, and at the high magnifications which are sometimes encountered in electron microscopy it will be huge (often more than ten metres). Microscopists should therefore experience little difficulty in positioning their viewing screen or photographic film.

1.8 Aberrations in optical systems

In discussing resolution and depth of field it has been assumed that all the components of the microscope are perfect and will focus the light from any point on the object to a similar unique point in the image. This is in fact rather difficult to achieve because of lens aberrations. The easiest lenses to make (for light) are those with spherical surfaces but any single spherical lens suffers from two types of aberration – *chromatic* aberrations which depend on the spectrum of wavelengths in the light and *monochromatic* or *achromatic* aberrations which affect even light of a single wavelength. The effect of each aberration is to distort the image of every point in the object in a particular way, leading to an overall loss of quality and resolution in the image. In order to correct these aberrations it is necessary to replace the single lens in the figures with compound lenses containing several carefully shaped pieces of glass with different refractive indices. Although this is not a correction technique which can be used in the electron microscope, the same types of aberration arise and are very important in determining the resolution of the instrument. Therefore the most significant aberrations must be considered in more detail.

Chromatic aberrations occur when a range of wavelengths is present in the light (e.g. in 'white' light) and arise because a single lens causes light to be deviated by an amount depending on its wavelength. Thus a lens will have different focal lengths for light of different wavelengths. To take an extreme example, a red focus and a blue focus will be formed from white light. Figure 1.12 makes this clear in terms of ray paths and illustrates that wherever the image is viewed a coloured halo will surround each detail. For example if the viewing screen is placed at A, the image of a point will appear as a 'bluish' dot with a red halo, whereas if the screen were at B the dot would be 'reddish' with a blue halo. In neither case is a truly focused image of a small white dot

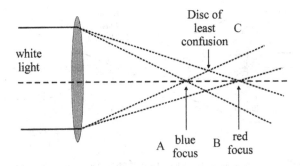

Figure 1.12 Ray diagram illustrating the introduction of chromatic aberration by a single lens. Light of shorter wavelength (blue) is brought to a focus nearer the lens than the longer wavelength (red) light. The smallest 'focused' spot is the disc of least confusion at C.

formed. With the screen at the compromise position C the smallest image is formed, but it is not a point but a *disc of least confusion*.

All aberration corrections are designed to reduce in size this disc of confusion. In the light microscope there are two ways in which chromatic aberrations can be improved, either by combining lenses of different shapes and refractive indices or by eliminating the variation in wavelength from the light source by the use of filters or special lamps. Both methods are often used if the very best resolution is required.

Monochromatic aberrations arise because of the different path lengths of different rays from an object point to the image point. The simplest of these effects is *spherical aberration*, which is illustrated in Figure 1.13. The portion of the lens furthest from the optical axis brings rays to a focus nearer the lens than does the central portion of the lens. Another way of expressing this concept is to say that the optical ray path length from object point to focused image point should always be the same. This naturally implies that the focus for marginal

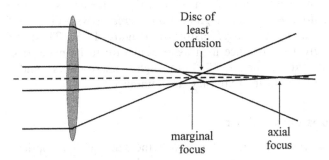

Figure 1.13 Ray diagram illustrating spherical aberration. Marginal rays are brought to focus nearer the lens than near-axial rays.

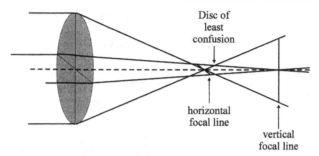

Figure 1.14 Ray diagram illustrating the formation of astigmatism for a lens with slightly different optical properties in the horizontal and vertical directions. In this illustration the lens is more powerful in the vertical plane.

rays is nearer to the lens than the focus for paraxial rays (those which are almost parallel to the axis). Again a disc of least confusion exists at the best compromise position of focus.

A related effect is that of *astigmatism*. For object points off the optical axis the path length criterion shows that there will be a focus for rays travelling in the horizontal plane at a different position from the focus for rays travelling in the vertical plane drawn in Figure 1.2(*a*). A similar, but more serious, effect occurs if a lens does not have identical properties across the whole of its area. As an example Figure 1.14 shows the effect for a lens with slightly different properties in the horizontal and vertical planes. All the monochromatic aberrations are reduced if only the central portion of the lens is used, i.e. if the lens aperture is 'stopped down'. Unfortunately, as explained in section 1.6, this limits the resolution of the microscope.

Other aberrations are often discussed in textbooks on optics but the three mentioned here are those of prime concern in electron microscopy. One further effect which is sometimes troublesome, particularly at low magnification, is distortion. This occurs if for some reason the magnification of the lens changes for rays off the optical axis. The two possible cases are when magnification increases with distance from the optical axis, leading to *pincushion* distortion, and when magnification decreases with distance from the optical axis, leading to *barrel* distortion (Figure 1.15). These effects are obviously of great importance if measurements are to be made from micrographs and manufacturers of both light microscopes and electron microscopes try to ensure that they are minimized.

1.9 Electrons versus light

In very many ways electron optics is just the same as light optics – all the terminology used in this chapter applies and ray diagrams can be used to illustrate the working of electron microscopes.

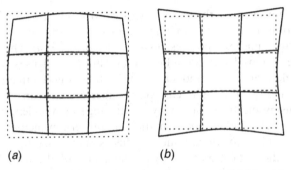

(a)   (b)

Figure 1.15 The appearance of a square grid in the presence of (a) barrel and (b) pincushion distortion.

For many purposes it is adequate to think of light as electromagnetic radiation with a wavelength λ and of electrons as sub-atomic particles. Both types of description (*wave* and *particle*) of course apply to both light and electrons: thus light may be described in terms of photons or as radiation of wavelength 400–700 nm, while electrons can also be considered as radiation with wavelengths (useful in microscopy) between about 0·001 and 0·01 nm. The first obvious difference between electrons and light is that their wavelengths differ by a factor of many thousands. The implications of this for microscopy are immense but fortunately in most cases lead to a simplification.

Another major difference is that electrons are very much more strongly scattered by gases than is light. This is so severe an effect that in order to use electrons in a microscope all the optical paths must be evacuated to a pressure of better than 10^{-10} Pa (about 10^{-7} of atmospheric pressure); the electrons would scarcely penetrate a few millimetres of air at atmospheric pressure.

Since, as will be shown in the next chapter, the lenses in an electron microscope are merely magnetic fields there is a negligible change of refractive index as the electrons pass through each lens. Hence in electron-optical calculations μ can be assumed to be unity. Furthermore the angles through which the 'rays' need to be deflected are generally very small (a few degrees) and the approximation $\sin\alpha = \tan\alpha = \alpha$ (Figure 1.10, in radians) holds to a very high degree of accuracy. These simplifications mean that the theoretical resolution of the electron microscope (equation 1.4) can be written as:

$$r_1 = \frac{0\cdot61\lambda}{\alpha} \tag{1.7}$$

which implies a resolution of about 0·02 nm, using reasonable values of $\lambda = 0\cdot0037$ nm (the wavelength of 100 kV electrons) and $\alpha = 0.1$ radians (about 5 degrees). This is much smaller than the size of a single atom.

Unfortunately, however, in the *transmission electron microscope* (TEM) this sort of resolution cannot be obtained because of the lens aberrations. Whereas in a light microscope it is possible to correct both chromatic and achromatic aberrations by using subtle combinations of lenses, this is very difficult using electron lenses and has only been seriously attempted in the 1990s. Consequently although chromatic aberrations can be virtually eliminated by using electrons of a very small range of wavelengths it is not possible to eliminate the monochromatic aberrations, principally spherical aberration. The only way of minimizing this is to restrict the electrons to paths very near the optical axis, i.e. near the centre of the lens, by using a small objective aperture. The importance of doing this can be seen from the equation for the degradation of resolution caused by spherical aberration in a lens of aberration coefficient, C_s.

$$r_2 = C_s \alpha^3 \tag{1.8}$$

The use of a small aperture thus reduces spherical aberration but makes the diffraction-limited resolution worse. There is an optimum size of aperture (i.e. value of α) for which the net resolution is smallest. This can be calculated quite easily by assuming that the net resolution is given by $r = r_1 + r_2$ and minimizing r with respect to α. The result is:

$$\alpha_{opt} = 0 \cdot 67 \lambda^{1/4} C_s^{-1/4}$$

$$r_{opt} = 1 \cdot 21 \lambda^{3/4} C_s^{1/4} \tag{1.9}$$

Under slightly different conditions it turns out that the resolution can be improved and the factor 1·21 can be reduced to as low as 0·7 in favourable circumstances. (Notice that a similar calculation is used, with an alternative derivation, to arrive at the resolution limit of the *scanning electron microscope* (SEM) in section 5.5.1.) Using the optimum aperture it is now possible with a good TEM to resolve two points about 0·2 nm apart. This is approximately the separation of atoms in a solid.

Since it is necessary to keep α small in order to reduce the effect of spherical aberrations, electron microscopes always gain the advantage of a large depth of field. Equation 1.5 can be re-written using the approximations appropriate to electrons as:

$$h = \frac{0 \cdot 61 \lambda}{\alpha^2} \tag{1.10}$$

which shows that as α is reduced the depth of field increases very rapidly. This is one of the major advantages of electron microscopy.

A further major difference between electrons and light is that electrons carry a charge. Not only does this mean that electromagnetic fields can be used as lenses for electrons but it opens up the possibility of easily scanning a beam of electrons back and forth, as happens in a cathode ray tube or a television tube.

The application of this approach has led to the development of the scanning electron microscope which, as is shown in Chapter 5, has over the past thirty-five years revolutionized attitudes to the study of surfaces.

With both types of electron microscope, transmission and scanning, the use of electromagnetic lenses and deflection coils means that it is possible to obtain an image of a specimen at any magnification within a wide range (say up to $1\,000\,000\times$), without physically changing or moving lenses. Electron microscopy therefore offers higher resolution, higher magnification, greater depth of field and greater versatility than the light microscope, albeit at a rather higher price.

1.10 Questions

1 If a small object is placed 2 mm away from a convex lens of focal length 1 mm, how far from the lens will the image be formed?

2 Where is the image produced by a thin convex lens when the object is at the focal point?

3 How many convex lenses, with focal length 1 mm and object distance (u) 1·1 mm, are needed to give a final image with magnification 1 million times?

4 In a light microscope an object is placed 2 mm away from a lens of diameter 2 mm. The object is in air (refractive index $= 1$) and the wavelength of the (green) light is 520 nm. What is the best possible resolving power of this microscope?

5 Calculate the position of the image and its magnification when an object is held 10 cm from a convex lens of focal length 8 cm.

6 Calculate the depth of field for a resolving power of 1 μm in a microscope with a final aperture of diameter 1 mm and a working distance of 20 mm. What is the depth of focus at a magnification of $100\times$?

7 If lenses with maximum useful magnifications of $40\times$ are available, how many lenses are needed to achieve magnifications of: $100\times$, $10\,000\times$, 1 million$\times$?

8 What are the dimensions of C_s? Deduce the approximate value of C_s for (a) an electron microscope capable of 0·1 nm resolution, and (b) a light microscope capable of 0·5 mm resolution.

9 'Chromatic aberrations can be virtually eliminated by using electrons of a very small range of wavelengths.' Why, in a TEM, can chromatic aberrations never be *completely* eliminated?

Chapter 2

Electrons and their interaction with the specimen

2.1 Introduction

When thinking about light microscopy we tend to ignore most of the interactions between the light and the specimen. It is sufficient that enough light is transmitted through or reflected from the specimen that the image can easily be seen. The assumption is generally made that the specimen is unchanged by the fact that it has been observed and for most specimens this is a reasonable assumption. However, the interaction of electrons with the material through which they pass may have more serious consequences. For example, as pointed out in Chapter 1, because of the scattering of electrons by gas molecules in the air, the column of an electron microscope must be evacuated. Some other real possibilities are that the specimen will be heated by the electron beam and that chemical changes may take place.

It is clearly important, in order to appreciate the way in which an electron microscope works and the meaning of the information which it provides, that we understand the nature of the possible interactions between the electron beam and the other parts of the microscope (e.g. lenses or camera) and between the electrons and the specimen. In order to achieve this understanding it is necessary to consider in more detail the nature of the electron and the various possible interactions between an electron and an atom. Those readers who need only a very superficial understanding of electron microscopy or who wish to get the 'feel' of the subject on a first reading may like to skip the rest of this chapter and the next and pass on to Chapter 4.

2.2 Electrons

Two schematic ways of looking at the structure of a typical isolated atom are shown in Figure 2.1. The nucleus carries a positive charge and is surrounded by a number of negative electrons which exactly neutralize this charge. When atoms are close to one another in a solid most of their electrons remain 'localized' – that is they can be considered to remain associated with a particular atom – but some outer electrons will be shared, to an extent which depends

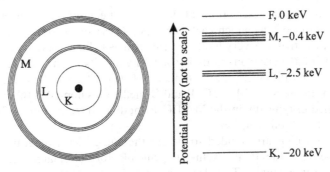

Figure 2.1 Two alternative representations of the first three electron shells around a molybdenum atom. The innermost (K shell) electrons are the most tightly bound, and they would need to be given approximately 20 keV before they could leave the atom.

on the type of bonding with neighbouring atoms. Several conventions have been developed to describe the states and energies of localized electrons – two of the most widespread will be described here.

It is common to define the zero of the energy scale as the potential energy of a free electron far from any atom, F. The energies of localized electrons are then negative, as shown in Figure 2.1. Alternatively, spectroscopists refer to a positive 'binding energy' or the energy of the atom with the specified electron missing, which is the negative of the energy shown in the figure. Table 2.1 gives

Table 2.1 Electron states.

| Shell | Sub-shell | | Possible number of electrons |
	KLM	spdf	
K	K	Is	2
L	LI	2s	2
	L2	2p	2
	L3	2p	4
M	MI	3s	2
	M2	3p	2
	M3	3p	4
	M4	3d	4
	M5	3d	6
N	NI	4s	2
	N2	4p	2
	N3	4p	4
	N4	4d	4
	N5	4d	6
	N6	4f	6
	N7	4f	8

the 'KLM' and 'spdf' descriptions of the 16 lowest energy states, together with the number of electrons which each can hold. These states are not necessarily occupied by electrons. A very light element such as helium, which only has two electrons, will have electrons in its K shell only and its L, M and higher shells will be empty. Uranium, with an atomic number of 92 and hence 92 electrons, has electrons in the K, L, M, N, O, P and Q shells. This nomenclature is defined and explained graphically in the MATTER module 'Introduction to Electrons in Crystals' (see bibliography).

For atoms which are bonded to others the energy level diagram must be modified because the Pauli exclusion principle forbids shared electrons to occupy the same states. This leads to the development of *energy bands*, the most common of which are referred to as the valence and conduction bands. In a metal the conduction band holds the familiar 'sea of electrons' which are responsible for conduction and much of the bonding. It is not necessary to consider atomic bonding in detail in order to understand electron microscopy, but the important terms which must be understood include:

Inner shell electrons: the lowest energy electrons, nearest to the nucleus, usually in the K or L shell. These electrons have very sharply defined energies, and are localized.

Outer electrons: the highest energy electrons (i.e. those with the lowest binding energies). In an isolated atom these will be in the outermost occupied shell.

Conduction band: shared outer electrons with a range of possible energies. These are delocalized.

Outer electrons are fairly readily detached from their atoms since only a small amount of energy need be supplied. It is this easy availability and low mass which makes 'free' electrons so useful. We must now consider the characteristics of a free electron in more detail.

An electron, considered as a particle, carries a single negative charge, e, of $1 \cdot 6 \times 10^{-19}$ C and has a rest mass, m_e of about 9×10^{-31} kg. If a single electron is accelerated through a potential difference V volts then its energy is V electron volts (eV). If V is large then its velocity, v, may well approach the velocity of light, c, and relativistic effects will become important. One of these is that its mass will increase according to the equation

$$m = \frac{m_e}{[1 - (v/c)^2]^{1/2}} \tag{2.1}$$

If the electron is now considered as a wave then its wavelength and momentum

Table 2.2 Electron Wavelengths.

V (kV)	Wavelength λ (nm) Uncorrected	Relativistically corrected
20	0·0086	0·0086
40	0·0061	0·0060
60	0·0050	0·0049
80	0·0043	0·0042
100	0·0039	0·0037
200	0·0027	0·0025
300	0·0022	0·0020
400	0·0019	0·0016
500	0·0017	0·0014
1000	0·0012	0·0009

$$c = 2\text{·}998 \times 10^8 \text{ ms}^{-1}$$
$$e = 1.602 \times 10^{-19} \text{ C}$$
$$h = 6.62 \times 10^{-39} \text{ Js}$$
$$m_e = 9.108 \times 10^{-31} \text{ kg}$$

are related by the de Broglie relationship

$$\lambda = \frac{h}{p} = \frac{h}{mv} \tag{2.2}$$

where h is the Planck constant. In addition the energy given to the electron, eV, can be equated to the energy represented by the relativistic change of mass, so

$$eV = (m - m_e)c^2 \tag{2.3}$$

Combination of equations 2.1, 2.2 and 2.3 shows that the wavelength of the electron depends on the potential difference, or *accelerating voltage*, in the following way:

$$\lambda^2 = h^2/(2eVm_e + e^2V^2/c^2)$$

which, on substitution of the values for h, e, m_e and c, becomes:

$$\lambda = [1\text{·}5/(V + 10^{-6}\,V^2)]^{1/2} \text{ nm} \tag{2.4}$$

At the accelerating voltages which are most useful for electron microscopy (2×10^4 V upwards) the electrons are accelerated to a velocity which is a significant fraction of the velocity of light and relativistic effects are quite important. Consequently the wavelength must be calculated according to equation 2.4 and not according to the common approximation $\lambda = (1\text{·}5/V)^{1/2}$ nm.

As Table 2.2 shows, the effect of relativity is quite marked for high accelerating voltages, amounting to a 25% correction at one million volts.

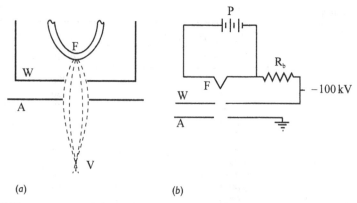

(a)   (b)

Figure 2.2 The geometry and electrical layout of a thermionic triode electron gun. (a) The electrons are emitted from a small region at the tip of a heated tungsten filament (F)
and are accelerated towards the anode (A). The fields generated between the filament and the anode, modified by the Wehnelt cylinder (W) acting as a grid, cause
the electrons to be focused at V, which is known as the virtual source. (b) The filament is heated by the passage of a current from the power supply P and the voltage
on the grid is determined by the bias resistor R_b.

2.3 Generating a beam of electrons

Of the many ways of encouraging electrons to leave a solid so that they may be
accelerated towards the specimen, two have proved particularly useful in the
construction of *electron guns*. The most widespread system uses thermionic
emission from a heated filament. At temperatures in excess of 2700 K, for
instance, a tungsten wire emits an abundance of both light and electrons; in
a light bulb only the light is used but in an electron gun the electrons are
accelerated across a potential difference of tens or hundreds of kilovolts to
generate a beam of electrons of controlled energy (and hence of known wavelength). The general features of a thermionic triode electron gun are shown in
Figure 2.2. Notice that a small current flows when the potential is applied but
before the filament current is high enough to give rise to thermionic emission.
This is known as the dark current, I_{dark}, since it flows before the filament is hot
enough to emit light.

A piece of tungsten, usually a wire bent into a hairpin, acts as the cathode.
This filament (F) is heated by the passage of a current to about 2800 K while
being held at a high negative potential with respect to the anode (A) and the
rest of the microscope. Electrons thermionically emitted from the filament are
accelerated rapidly towards the anode and a beam of high energy electrons is
emitted through the circular hole at its centre into the microscope column. The
addition of a Wehnelt cap (W), which is held at a voltage slightly more negative
than the filament, enables the diameter of the area at the end of the filament
which emits electrons to be controlled. The Wehnelt cap acts rather like the
grid in a triode valve (or the base in a bipolar transistor) and hence this gun is

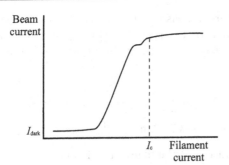

Figure 2.3 The electron emission curve for a triode gun. As the filament current is increased the beam current rises steeply but then saturates when the filament current reaches I_c.

usually called a triode gun. An important feature of the gun is that the paths of the emitted electrons usually cross at one point in space and the gun acts as a lens. The diameter of the beam at the crossover is dependent on the area of the filament which is emitting electrons and this can be controlled by the difference in potential between the filament and the grid (i.e. by R_b, the value of the bias resistor in Figure 2.2). The crossover diameter, analogous to the disc of least confusion which was defined in Chapter 1, is effectively the size of the electron source, and is of great importance in calculating the resolution of a scanning electron microscope.

The triode circuit shown in Figure 2.2(*b*) acts rather subtly to limit the current in the emitted electron beam. As the emitted current increases, so does the bias voltage and this suppresses further electron emission. This is known as the autobias mechanism and accounts for the characteristic shape of the emission curve shown in Figure 2.3. As the current through the filament is increased there is an initial rise in the emitted electron beam current. This eventually saturates however and there is no point in passing more than the critical current I_c through the filament since this merely increases the temperature of the filament (thus reducing its lifetime) without giving rise to any additional beam current. The operation of a triode gun is simulated in the MATTER module 'Introduction to Electron Microscopes' (see bibliography). This can be used to experiment with the effect of filament current on the emission current.

The thermionic gun is satisfactory for many purposes but is limited in the *brightness* of the beam it can produce. In the context of electron microscopy brightness is defined as the beam current density per unit solid angle, usually cited in units of A m^{-2} sr^{-1}.*

---

*   A steradian (sr) is the unit of solid angle. One steradian is defined as that solid angle which encloses a surface on a sphere equal to the square of its radius. Therefore 4π steradians occupy the whole of space.

Brightness is therefore a measure of how many electrons per second can be directed at a given area of the specimen. For a beam of current density j_c ($\mathrm{A\,m^{-2}}$) with a convergence angle β, the brightness, B, is thus given by:

$$B = j_c/\pi\beta^2$$

For a thermionic gun B is limited to

$$B = 2 \times 10^5\, TV \exp(-\phi/kT)\,\mathrm{Am^{-2}\,sr^{-1}} \tag{2.5}$$

where T is the temperature of the filament in K and ϕ is the thermionic work function of the filament material in electron volts. Since B increases rapidly as T increases and as ϕ decreases, it is best to use a filament material with as high a melting point and as low a work function as possible. Tungsten has a high melting point (3653 K) and a work function which is much the same as most metals (4·5 eV) and is the most widely used filament material. Tungsten filaments give a brightness of about $10^9\,\mathrm{Am^{-2}\,sr^{-1}}$. However, the brightness can be increased by a factor of 10 or more if LaB_6, with a work function of 3·0 eV, is used. Electron guns which use LaB_6, are quite common on microscopes which are used for analytical or high resolution work since in both of these fields a high brightness is desirable.

If still higher brightness is required then the *field emission gun* is used. If a metal surface is subjected to an extremely high electric field ($> 10^9$ V/m) there is a high probability that an electron can leave the surface without needing to be given the amount of energy represented by the work function. This is because the effect predicted by quantum mechanics and known as *tunnelling* can occur. The result is that many more electrons can be drawn from a piece of tungsten than is possible using thermionic emission and the brightness can be increased by a factor of a thousand or more to a value in excess of $10^{13}\,\mathrm{Am^{-2}\,sr^{-1}}$. The field emission current depends very strongly on the applied field, F, according to the Fowler–Nordheim equation;

$$j = 6\cdot2 \times 10^{-6}\frac{(E_f/\phi)^{1/2}\,F^2}{(E_f + \phi)}\exp\left(\frac{-6\cdot8 \times 10^9\,\phi^{3/2}}{F}\right)\mathrm{Am^{-2}}$$

where E_f is the Fermi energy, which is about 5 eV for tungsten at room temperature. For a field in excess of about $5 \times 10^9\,\mathrm{Vm^{-1}}$ the current emitted by field emission at room temperature exceeds that which can be thermionically emitted. In order to apply such a high field the emitter, usually tungsten, has to be prepared in the form of a sharp point. The diameter at the point must be about 0·1 µm, which is orders of magnitude finer than a pin, so the emitter is a rather delicate structure. For this fine point to be preserved in use it must be operated in an environment with very few ions and this dictates the use of ultra-high-vacuum (UHV) techniques. The vacuum in the gun must be lower than 10^{-7} Pa, which is rather better (and more expensive to achieve) than the value of 10^{-2} or 10^{-3} Pa which is common in thermionic guns. The gun

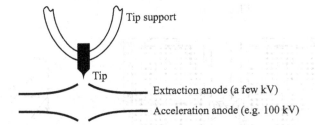

Figure 2.4 A field emission gun. Electrons are extracted from the fine tip by a first anode and then accelerated down the column by a second anode at a much higher potential.

configuration is rather different from the thermionic gun and is shown in Figure 2.4.

An important feature of field emission sources is that the emitted electrons have very well defined energies. Whereas the electrons from a thermionic source inevitably have an energy spread of 1–2 eV, the electrons from a cold field emission gun have a much smaller energy spread, usually less than 0·5 eV. This is particularly important, as will be shown in Chapter 6, for the interpretation of electron energy loss spectra (EELS). Thus for both analytical and high resolution electron microscopy, field emission sources are important not just because they provide a high brightness beam but because they provide a 'clean' monochromatic supply of electrons.

2.4 Deflection of electrons – magnetic lenses

It was realized in the 1920s that a beam of electrons could be focused by either an electrostatic or a magnetic field. Both types of field have been used in electron lenses but the electromagnetic lens is by now virtually universal in commercial electron microscopes so we will not consider the earlier electrostatic lenses any further.

The key to an understanding of what is essentially a very simple lens is the direction of the force which acts on a moving electron in a magnetic field. If an electron moving with velocity v experiences a magnetic field of strength B, then it suffers a force of $F = Bev$ in a direction perpendicular to both the direction of motion and the magnetic field. Expressed more concisely in vector notation:

$$F = e(B \wedge v)$$

A typical electromagnetic lens is designed to provide a magnetic field almost parallel to the direction of travel of the electrons. An electron entering the lens (Figure 2.5) experiences a magnetic field B which can conveniently be resolved into components B_{ax} along the axis of the microscope and B_{rad} in a radial direction. Initially the electron is unaffected by B_{ax}, which is parallel to its direction of travel, but experiences a small force of magnitude $B_{rad} ev$ from

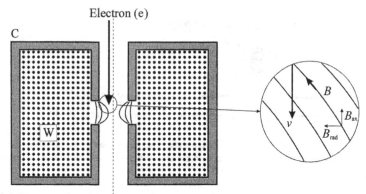

Figure 2.5 The action of a typical electromagnetic lens. An electron (e) entering the lens just off the optical axis (dotted) experiences a magnetic field such as that shown in the inset.

the small radial component. This force causes the electron to travel in a helix along the lens. As soon as it starts to spiral it has a component of velocity v_{circum} perpendicular to the plane of the paper and therefore experiences a force of magnitude $B_{ax} ev_{circum}$ in a radial direction. Thus the helical path follows a tighter and tighter radius and the effect is that a parallel beam of electrons entering the lens is caused to converge to a point exactly as light is focused by a glass lens. If the magnetic field only extends over a short distance along the axis, then the lens behaves as a 'thin lens' and all the geometrical expressions quoted in Chapter 1 apply.

In order to generate a magnetic field of just the right strength, size and shape, an arrangement similar to that shown in Figure 2.5 is almost invariably used. A coil consisting of a large number of turns of wire, W, is wound on a soft iron core, C (the pole piece), which has only a very small accurately machined air gap, G, across which the field is produced. By varying the current passing through the coil, typically in the range 0–1 A, the magnetic field strength and hence the focal length of the lens can be varied at will.

Although the several lenses in any one electron microscope may differ in shape and size they will conform to the general pattern laid out in Figure 2.5. An important feature, for which there is no analogy in the light microscope, is the spiralling of the electrons as they travel through an electromagnetic lens. Since it is very rare for the electron to travel an integral number of turns of the spiral as it passes through the lens, in general there is a rotation of the image caused by the lens. This is not a distortion, since the image is otherwise unaffected, but it does lead to one or two effects, particularly in transmission electron microscopes, which need to be borne in mind when looking at micrographs and electron diffraction patterns in later chapters. Cunning design of electron-optical systems in modern microscopes sometimes involves using lenses in combinations which cancel out the image rotation so that the effect is not apparent, although it is still present.

Electromagnetic fields are also used to deflect the entire beam of electrons, for example to scan the beam back and forth in a scanning electron microscope. For these applications the field needs to be perpendicular to the electron beam, but much smaller fields suffice for these smaller deflections so the coils are quite small. Most microscopes will contain a dozen or more such coils, designed to enable the microscopist to optimize the position of the beam in the column of the microscope.

The working of a typical lens, together with the operation of lenses and deflection coils, is demonstrated in the MATTER module 'Introduction to Electron Microscopes'.

2.5 The scattering of electrons by atoms

In almost all types of electron microscope *primary* electrons enter the specimen and the same or different electrons leave it again to form the image. Consequently it is vitally important to understand the interactions which are possible between high energy electrons and the atoms of the specimen. Without this understanding it is not possible to interpret the image, diffraction pattern or analytical spectrum which each type of microscope produces.

There is a set of terminology common to all electron scattering and we will consider this first. The probability that a particular electron will be scattered in a particular way is usually described in terms either of a *cross-section*, σ, or of a *mean free path*, λ. A cross-section is expressed as the area which the scattering particle *appears* to present to the electron. If there are N particles per unit volume of the specimen and the cross-section for a particular scattering event is (then the probability of a single electron being scattered in this way in its passage through a thickness dx of the specimen is $N\sigma dx$. An alternative way of expressing the same idea is to define the mean free path for the scattering as

$$\lambda = 1/N\sigma \qquad (2.6)$$

λ has the dimension length and is effectively the *average* distance which an electron will travel before being scattered in the specified way.

Mean free paths for many scattering processes are similar to the thickness of a TEM specimen. This means that electrons will tend to be scattered either once or not at all while passing through a thin specimen. On the other hand if an electron is incident on a thick specimen (e.g. in an SEM) it will be scattered many times until it effectively comes to rest. The terms *single scattering*, *plural scattering* and *multiple scattering* are often used to describe the situations in which electrons are scattered no more than once, several times and many times respectively.

In single or plural scattering situations the probability of an incident electron suffering n scattering events while travelling a distance x is given by the Poisson equation

$$p(n) = (1/n!)(x/\lambda)^n \exp(-x/\lambda) \tag{2.7}$$

Thus the probability of an electron undergoing exactly two scattering events with a mean free path λ in a distance t is $p(2)$ and is given by

$$p(2) = (1/2)(t/\lambda)^2 \exp(-t/\lambda)$$

Notice that $p(0)$ is the probability of the electron not being scattered by this process and that $1 - p(0)$ is therefore the probability of it being scattered once *or more*.

The Poisson equation approach is not much use for multiple scattering, where all primary electrons can be assumed to be scattered very many times, possibly by several different mechanisms. In these cases other averaging approaches are more fruitful; one example, the Monte Carlo method, is illustrated in section 2.7.

2.6 Elastic scattering

Elastic scattering is defined as a process which, although it might change the direction of the primary electron, does not change its energy detectably. This type of scattering results from Coulombic interactions (i.e. involving electrostatic charges) between the primary electron and both the nucleus and all the electrons around it. This is known as Rutherford scattering and it gives rise to a strongly 'forward peaked' distribution of scattered electrons. If the energy of the primary electron is E_0, the probability $p(\theta)$ of it being scattered through an angle θ is given by

$$p(\theta) \propto \frac{1}{E_0^2 \sin^4 \theta} \tag{2.8}$$

The probability of a small angle of scatter is very much greater than that of a large angle. Notice too that the probability of scattering through any angle decreases as the energy of the electron increases.

The mean free path for elastic scattering depends quite strongly on the atomic number of the scattering atom. To give an example for 100 kV electrons, λ is about 5 nm for gold (atomic number $Z = 79$) but about 150 nm for carbon ($Z = 6$).

Elastic scattering is important in electron microscopy because it is a major mechanism by which electrons are deflected and also because elastically scattered electrons are the main contributors to diffraction patterns. Most of Chapter 3 thus refers implicitly to elastic scattering. The strength of scattering by an atom depends on its atomic number and is usually described in terms of its atomic scattering factor, f. This is defined as the amplitude of scattering from the atom divided by the amplitude of scattering from a single electron.

2.7 Inelastic scattering

Inelastic scattering is a very general term which refers to any process which causes the primary electron to lose a detectable amount of energy, ΔE. In terms of the facilities usually available to electron microscopists, ΔE would need to be substantially more than 0·1 eV before it could be detected. There are many interaction processes which could cause energy to be lost by the primary electron and transferred to the electrons or atoms of the specimen. We will only consider four of the most probable types of scattering event. It is important to realize that the inelastic scattering processes (probably in combination) are eventually responsible for the stopping of an electron by a solid. Almost all of the kinetic energy which was carried by the primary electron will end up as heat in the specimen. A small proportion of the energy may escape as X-rays, light or secondary electrons and these may prove extremely useful for both imaging and analysis, as we show in Chapters 5 and 6. Secondary effects are dealt with in the next section; first let us consider the main types of inelastic scattering process.

2.7.1 Phonon scattering

Phonons are the quanta of elastic waves, that is of atomic vibrations in a solid. A primary electron can lose energy by exciting a phonon and effectively heating the solid slightly. The amount of energy lost in doing this is rather small, generally less than 1 eV, and the mean free path for high energy electrons is quite large, of the order of μm. These facts do not mean that phonon scattering is unimportant, for two main reasons. All electrons which remain in the solid are likely to excite phonons eventually, perhaps after they have lost larger amounts of energy by other means (see below) and this is how the solid is heated by the electron beam. Also, when phonon scattering occurs the scattered electron is generally deflected through quite a large angle, typically 10 degrees. This will be significant in the discussion of image contrast in Chapter 4.

2.7.2 Plasmon scattering

A plasmon is a wave in the 'sea' of electrons in the conduction band of a metal. There are similar effects among the bonding electrons of non-metals. In exciting a plasmon the primary beam loses 5–30 eV and the mean free path for this event is short – a few hundred nm in most materials. Consequently plasmon scattering is a frequent occurrence in all electron–solid interactions. We shall see in Chapter 6 that these energy losses dominate the 'energy loss spectrum' but are not very useful for analysis because the energy loss is not particularly characteristic of the scattering element. The Poisson equation (2.7) can be used to calculate how many plasmons an electron is likely to excite in traversing a thin specimen.

2.7.3 Single valence electron excitation

It is possible, but less likely, that a primary electron will transfer some energy to a single valence electron rather than to the 'sea' collectively. The mean free path for this process is quite large (μm), the energy loss is small (about 1 eV) and the typical scattering angle is also small so the process is not exploited in electron microscopy.

2.7.4 Inner shell excitation

A rare but valuable form of inelastic scattering is the knocking out of an inner shell electron. Because the binding energy of such K and L shell electrons may be high the energy lost by the primary electron can be quite large. For example it takes 283 eV to remove a carbon K electron, 69 508 eV to knock out a tungsten K and 1100 eV to excite a copper L. The mean free path for this type of scattering is quite large (μm), so the process occurs much less frequently than, say, plasmon scattering. However the secondary effects produced when the excited atom relaxes are almost ideal for analysis, as the next section shows. The cross-section for inner shell excitation, in common with that for most scattering processes, drops as the primary electron energy E_0 increases. It is also lower for elements of higher atomic number Z, since E_c (the critical energy to excite an X-ray) increases with Z and

$$\sigma \propto 1/E_c E_0 \tag{2.9}$$

2.7.5 Inelastic scattering and absorption

In all but the thinnest specimen it is clear that more than one of the inelastic scattering processes can take place. In a 'solid' specimen many such events will occur until the electron is stopped or leaves by the surface it entered. The trajectories of a few typical electrons, calculated by a Monte Carlo method which introduces the angles of scattering with their appropriate probabilities, are shown in Figure 2.6. It can be seen that the majority of electrons are brought to a halt within the solid but a few are *backscattered* and leave the specimen. The volume within which 95% or so of the primary electrons are brought to rest is generally referred to as the *interaction volume* and will be discussed in greater detail in Chapters 5 and 6. An interactive Monte Carlo simulation is available in the MATTER software module 'Electron Beam – Specimen Interactions'.

In the case of a solid specimen it is clear what is meant by *absorption*. For a thin specimen, through which many or most of the electrons are eventually transmitted, the term has to be defined very carefully. The geometry of an electron detection system is typically like that shown in Figure 2.7. By 'absorbed' we mean 'not detected in our experiment' and for this geometry that means 'scattered through an angle greater than θ'. If the atomic

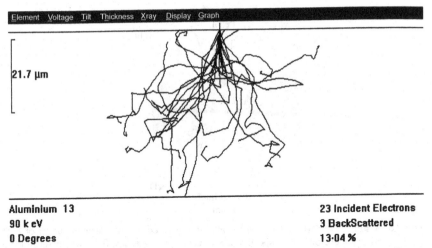

Element Voltage Tilt Thickness Xray Display Graph

21.7 µm

Aluminium 13	23 Incident Electrons
90 k eV	3 BackScattered
0 Degrees	13·04 %

Figure 2.6 Electron trajectories in aluminium calculated by a Monte Carlo procedure. Twenty three electrons are shown in this figure, of which three escape and are thus back-scattered.

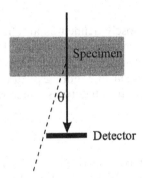

Figure 2.7 The geometry of an electron detection system. Electrons which are deflected through an angle greater than θ after passing through the specimen are not registered by the detector and have thus effectively been 'absorbed' in this experiment.

cross-section for scattering through θ or more is σ_a then it is possible to describe absorption using the conventional Lambert–Beer law. The equation is:

$$I/I_0 = \exp(-N\sigma_a x) \qquad (2.10)$$

This describes the fractional intensity, I/I_0 which remains after absorption in a thickness x. N is the number of scattering atoms per unit volume which is often convenient to write in terms of Avogadro's number, N_A, as $N = N_A\, \rho/A$ where ρ is the density and A the atomic mass.

Remember that all scattering mechanisms, elastic and inelastic, are strongly forward peaked so that the cross-section which is appropriate in equation 2.10 will depend quite sensitively on the angle θ subtended by the detector.

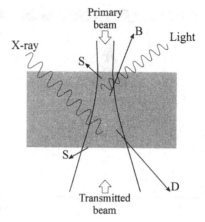

Figure 2.8 A summary of the effects which may be detected when a primary beam of high energy electrons hits a specimen. S = secondary electrons; B = backscattered electrons; D = diffracted electrons.

2.8 Secondary effects

A secondary effect can be loosely defined as an effect caused by the primary beam which can be detected outside the specimen. The secondary effects with which we will be mainly concerned are either electrons or electromagnetic radiation. They are summarized in Figure 2.8. For the present purpose it is convenient to categorize them into five types which relate to the way in which each effect is exploited.

2.8.1 Secondary electrons

This is not a very accurate term, but is used to describe those electrons which escape from the specimen with energies below about 50 eV. They could conceivably be primary electrons which at the very end of their trajectory (Figure 2.6) reach the surface with a few eV remaining. However they are more likely to be electrons to which a small amount of energy has been transferred (by one of the processes outlined in section 2.7) within a short distance of the surface. The *yield* of secondary electrons, that is the number emitted per primary electron, can be as high as, or higher than, 1. Secondary electrons are therefore abundant and are the most commonly used imaging signal in scanning electron microscopy (Chapter 5).

2.8.2 Backscattered electrons

It has already been shown in Figure 2.6 that some primary electrons may leave the surface before giving up all their energy. It turns out that they are most likely to do this while they still have a large fraction of their incident energy. Backscattered electrons are not usually as numerous as secondary electrons but most of them carry high energies. They are used for imaging, diffraction and analysis in the SEM (Chapter 5).

2.8.3 Relaxation of excited atoms

If a localized electron has been knocked out of an atom the atom is in an excited, high energy, state. At some later time the empty electron state will be filled and the atom will relax, giving off the excess energy as a secondary effect. There are essentially three ways in which this relaxation can happen. If the vacant electron state is an outer state then the energy to be given off will be small and is commonly emitted in the form of a photon which may be in the visible range. This effect is known as *cathodoluminescence*.

If on the other hand the vacant state is an inner state the amount of energy to be released is larger and there are two main possibilities: a characteristic X-ray or a characteristic (Auger) electron may be emitted. These two processes are shown schematically in Figure 2.9. If an X-ray is to be emitted a single outer electron jumps into the inner shell vacancy (Figure 2.9(a)). The energy of the X-ray is then the difference between the energies of the two excited states and this is characteristic of the particular atomic species. For example if a K shell electron has been knocked out of a molybdenum atom (Figure 2.1), and replaced by the 'jumping in' of an L electron the energy difference, ΔE, is 17 400 eV, which is emitted as the K_α X-ray of Mo. The wavelength of this X-ray can be calculated from

$$\lambda = \frac{hc}{\Delta E} \qquad (2.11)$$

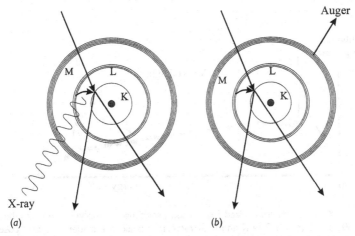

Figure 2.9 The two ways in which an inner-shell-excited atom can relax. In both cases a K shell electron has been knocked out – in (a) a characteristic (K_α) X-ray is emitted while in (b) an Auger electron (KLM) is ejected.

and in this case is 0·071 nm. If the vacancy had been filled by an M shell electron the energy difference would have been greater, 19 600 eV, and the X-ray emitted would have been the Mo K_β, with a wavelength of 0·063 nm. These energies and wavelengths are different for each atomic species and by measuring them we can determine which elements must have been present in the specimen. This is the basis of analytical electron microscopy and electron probe microanalysis (Chapter 6).

Inspection of Table 2.1 would lead to the conclusion that there should be very many characteristic X-rays for each atom, since transitions between all the possible states for each atom would appear to be possible. However a set of selection rules prohibit many of the transitions, so that for instance a 2s electron cannot jump to fill a 1s vacancy. The most useful X-rays are listed in Chapter 6.

It is possible for a primary electron to excite an X-ray without knocking out an inner shell electron. In this case the electron can lose any amount of energy (up to its total kinetic energy) and the X-ray is no longer characteristic of a particular atom. This process is called Bremsstrahlung (German for 'braking radiation') and leads to a background of X-rays in any electron-generated X-ray spectrum. Figure 2.10 shows a typical spectrum with both characteristic and Bremsstrahlung radiation.

An alternative to X-ray emission is the ejection of an outer electron carrying the excess energy as kinetic energy. This process, known as Auger emission, is shown in Figure 2.9(b). Three electrons are now involved; the original vacancy, the outer electron which jumps into it and the other outer electron which leaves carrying the surplus energy. Measurement of the energy of the characteristic Auger electrons forms the basis of *Auger electron spectroscopy*, which is discussed in Chapter 7.

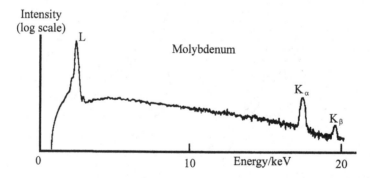

Figure 2.10 An X-ray spectrum excited from a solid specimen of molybdenum by a beam of 30 keV electrons. The K and L characteristic peaks are evident, superimposed on the Bremsstrahlung background. Notice that the intensity scale is logarithmic so that the background intensity appears to be rather higher in comparison with the characteristic peaks than it is in reality.

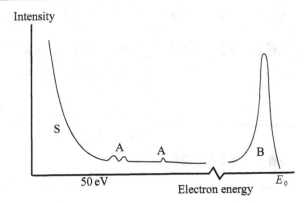

Figure 2.11 An idealized electron spectrum showing the relative abundance of secondary (S), Auger (A) and backscattered (B) electrons. Notice that the energy scale is not continuous; E_0 is typically $\gg 50\,$eV.

Auger electron and characteristic X-ray emission are alternative processes by which energy can be emitted as an excited atom relaxes. The two processes do not, however, occur with equal probability and the fraction of atoms which emits an X-ray, known as the fluorescence yield, varies strongly with atomic number. A simple expression for the X-ray fluorescence yield, w, is

$$w = Z^4/(Z^4 + c)$$

Z is the atomic number and the constant c has a value of about 10^6 for atoms whose K shells have been excited and is larger for the L and M shells. The yield of Auger electrons is $1 - w$. A simple calculation shows that from a light element (small Z) Auger electrons will be emitted in far larger numbers than X-rays while for heavy elements the situation is reversed.

The relative abundance of the three types of electron emission (secondary, backscattered and Auger) is illustrated in Figure 2.11, which shows an idealized electron spectrum. There are large numbers of secondary and backscattered electrons but relatively few of the (analytically most useful) Auger electrons.

The mechanisms of X-ray and Auger emission are described graphically in the MATTER module 'Introduction to Electrons in Crystals'. This can be used to show dynamically the electrons sketched in Figure 2.9.

2.9 The family of electron microscopes

The ideas presented in this chapter contain all the elements needed to construct the whole family of electron microscopes. All instruments have an electron gun, a system of condenser lenses, and some sort of signal detector. Figure 2.12 shows the way in which these and other components are put together to make the various types of EM which are in common use. Starting in the centre of the diagram, the SEM needs scanning coils and an electron detector, while

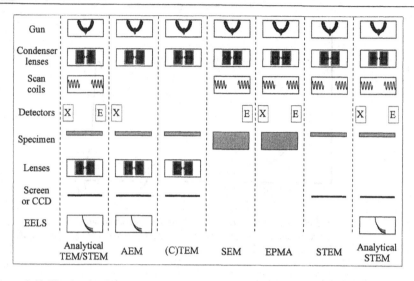

Figure 2.12 The family of electron microscopes. The two basic imaging systems (SEM and TEM) are shown in the centre. The addition of further detectors or lenses can give each microscope a range of analytical capabilities. $X = $ X-ray detector; $E = $ electron detector.

the TEM needs objective and projector lenses and a viewing screen. From these two basic arrangements all the other microscopes to be described in this book have developed. The SEM becomes a microprobe analyser (EPMA) when an X-ray detector is added, or a STEM when the detector is moved below the specimen. In a similar way an X-ray detector and/or an electron energy loss spectrometer (EELS) can be added to a TEM to make an analytical EM (AEM). Addition of scanning coils to this configuration results in a TEM/STEM, while adding post-specimen lenses to a dedicated STEM at the other end of the diagram would achieve virtually the same configuration. The two ends of Figure 2.12 therefore effectively meet. All the instruments illustrated in the figure are described in more detail in later chapters.

2.10 Questions

1. An electron beam of brightness (B) $10^{10}\,\mathrm{Am^{-2}\,sr^{-1}}$ is focused to a spot of diameter 100 nm at the specimen. What is the current density ($\mathrm{Acm^{-2}}$) within the spot and what is the dose rate in electrons per square nm per second? Take the convergence angle to be $0 \cdot 04\,\mathrm{rad}$.

2. By considering the X-ray fluorescence yield for K shell excitation deduce which element should give almost equal numbers of characteristic X-rays and Auger electrons.

3. What is the increase in mass (above the rest mass) of a 300 kV electron?

4   Would you expect the energy of an Auger peak to be higher, lower or the same as that of the equivalent X-ray peak? Explain why.

5   What fraction of electrons will not be scattered in a thickness of 100 nm if the mean free path for all scattering is 80 nm? What fraction will be scattered three times? [Use the Poisson distribution on page 30]

6   A detector set up as shown in Figure 2.6, subtending an angle 0·1 rad, detects 90% of the electrons passing through the specimen. If the detector is replaced by a larger detector, subtending 0·2 rad, what fraction of the electrons should now be detected? Assume that the scattering has an angular dependence given by the Rutherford equation (2.8).

7   Use the de Broglie relationship to deduce the effective wavelength of a pool ball of mass 0·1 kg travelling at 2 ms^{-1}.

8   The electronic configuration of aluminium (atomic number 13) is $1s^2 2s^2 2p^6 3s^2 3p^1$. What is the configuration of copper (atomic number 29)?

9   Calculate the wavelength of an X-ray of energy 2166 eV. Deduce from the table of wavelengths in Chapter 6 (Table 6.1) which elements it might have arisen from.

10 What advantages and disadvantages would you expect if the electron gun of a TEM was located at the bottom of the column rather than in its usual position at the top?

Chapter 3

Electron diffraction

As we saw in the previous chapter, an electron beam which has passed through a thin specimen contains two components: elastically scattered electrons and inelastically scattered electrons. From an analysis of the spatial distribution of the scattered electrons – known as an electron diffraction pattern, we can deduce a great deal of information about the arrangement of the atoms in the specimen.

Equation 2.8 shows that the intensity of elastic scattering (f_θ) from a particular atom species is a maximum when the scattering angle (θ) is zero and decreases monotonically as θ increases. We might therefore expect that a diffraction pattern from a solid specimen would be somewhat similar. Figure 3.1(a) is a diffraction pattern from a thin film of amorphous carbon, and Figure 3.1(b) is a graph of intensity as a function of scattering angle. It is clear that although the intensity of the scattering does generally decrease as θ increases, the variation of scattered intensity is more complex than equation 2.8 would predict as there are some diffuse maxima in the intensity. As may be seen from Figures 3.1(c) and 3.1(d), the diffraction patterns from crystalline materials are even more complex. Although *individual atoms* scatter according to equation 2.8, when the atoms are close together, there is a strong interaction between the electrons scattered from different atoms. Therefore, in order to account for the diffraction patterns from solids, we must look in some detail at the effects of atomic arrangement on electron scattering. Although some information about the average atomic spacing in an amorphous material may be obtained from patterns such as Figure 3.1(a), it is from crystalline materials that we can obtain the most information, and the rest of this chapter will deal exclusively with crystalline materials.

The two parameters of importance in electron diffraction are the *angular distribution* of the scattered electrons and the *intensity* of the scattering. The geometry of electron diffraction patterns is fairly simple, and from a knowledge of this we can deduce a great deal of information about the structure of a crystal and its orientation. A knowledge of the factors which determine the intensity of electron scattering enables us to derive more detailed information from a diffraction pattern, and, perhaps more importantly, enables us to understand and interpret the images of crystalline materials in the transmission

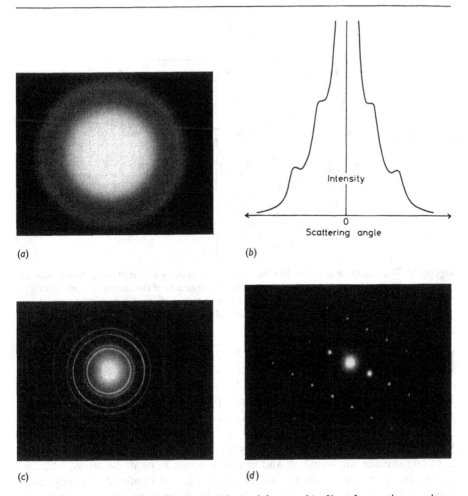

Figure 3.1 (a) Electron diffraction pattern obtained from a thin film of amorphous carbon.
(b) The variation of intensity with scattering angle obtained from Figure 3.1(a).
(c) Diffraction pattern from a fine grained polycrystalline gold specimen.
(d) Diffraction pattern of a single crystal of aluminium.

electron microscope, as will be discussed in Chapter 4. However, a detailed calculation of the intensity of diffraction is beyond the scope of this book, and we shall only look at this aspect of diffraction briefly.

3.1 The geometry of electron diffraction

3.1.1 Bragg diffraction

Let us first consider the interaction of an electron beam with a perfect crystal in which all the atoms lie on a cubic lattice. A cross-section of a very thin

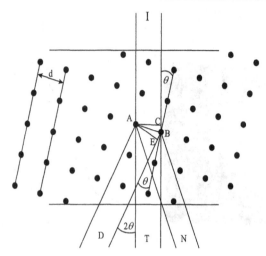

Figure 3.2 The scattering of an incident beam of electrons (I) by a crystalline specimen. Intense beams of electrons may emerge from the other side of the specimen undeviated (T) or having been diffracted (D) from atomic planes of spacing *d*. In other directions (e.g. N) no intense beams will be formed.

specimen of this material is shown in Figure 3.2. Now if an electron beam is incident on this specimen it will be elastically scattered by some of the constituent atoms such as those marked A and B. We can think of the electron beam as a wave motion (section 2.2) and apply a similar argument to that used to explain the diffraction of light or X-rays.

The incident electron beam is locally coherent; in other words, all the individual electron waves are in phase. Any scattered waves which are also in phase with one another will reinforce and lead to a strong beam of electrons, whereas any scattered waves which are out of phase will not reinforce. The scattered waves at D will be in phase if the path lengths differ by an integral number of wavelengths. This will be so if $CB + BE = n\lambda$, where n is an integer and λ is the electron wavelength. However, it may be seen from Figure 3.2 that $BE = CB = d \sin \theta$, and hence the condition for reinforcement is

$$2d \sin \theta = n\lambda. \tag{3.1}$$

This is well known as *Bragg's law* and it is widely applied to the diffraction of X-rays as well as electrons. It tells us that we can expect very few elastically scattered electrons to emerge from our specimen unless they are at an angle θ which is a solution of equation 3.1.

We need to consider the significance of d, n and θ. As may be seen from Figure 3.2, d is the spacing of the atoms which are scattering the electrons. In a crystalline array, d is therefore the spacing of the lines of atoms, or in a three-dimensional crystal, the spacing of planes of atoms – the *interplanar spacing*. In any crystal there will be many other planes which we can define, each of which will have its own interplanar spacing.

In order to interpret real diffraction patterns we need a formal system of defining planes and directions in the crystal, and we will use the standard *Miller index notation*, details of which may be found in any textbook dealing with elementary crystallography. Briefly, in this notation a direction in the crystal is denoted by square brackets, and is expressed as three numbers (usually integers) which give the co-ordinates of the direction using the *x*, *y* and *z* axes of the unit cell, e.g. [110]. A plane in the crystal is denoted by round brackets, and is again expressed as three numbers which give the inverse of the distance along each crystal axis where the plane intercepts it, e.g. (110). Finally, since each strongly diffracted beam can be associated with a particular set of crystal planes, diffracted beams are also indexed using Miller indices, but in this case there are no brackets, e.g. 110. In all cases a negative number is denoted by a bar above, to save space, e.g. ($\bar{1}$10). Using this notation for a cubic crystal for example, the interplanar spacing of planes with the indices (*hkl*) is given by

$$d_{hkl} = \frac{a}{\sqrt{h^2 + k^2 + l^2}} \qquad (3.2)$$

where *a* is the length of an edge of the smallest cube which can be obtained with atoms at all corners (the *unit cell*). The integer *n* in equation 3.1 is the *order of diffraction*, and for a particular plane, diffraction occurs when $n = 1, 2, 3 \ldots$ etc. However, in electron diffraction it is conventional to consider only the first order of diffraction, i.e. $n = 1$, and to deal with higher orders by using the corresponding multiples of the Miller indices. Thus when dealing with the second order diffraction from a (131) plane, we imagine planes to be present with half the spacing and call this the first order diffraction from a (262) plane. Equation 3.1 is then written as

$$\lambda = 2d \sin \theta. \qquad (3.3)$$

Now let us consider the significance of the angle θ, the *Bragg angle*. Because of the very short wavelength of the electrons used in electron microscopy we can simplify equation 3.3 even further. If we substitute typical values of λ and d in equation 3.2, say $\lambda = 0.0037$ nm (the value for 100 kV electrons from Table 2.2) and $d = 0.4$ nm (the spacing of some commonly occurring planes in aluminium), we find that $\sin \theta = 0.0046$ and $\theta = 0.26° = 0.0046$ radians. For such an angle, which is typical of electron diffraction, we can write $\sin \theta = \theta$, and hence equation 3.3 becomes:

$$\lambda = 2d\theta. \qquad (3.4)$$

Since θ is so small, it is obvious that we have exaggerated the angles in Figure 3.2 and that in practice, *there will only be strong diffraction from planes of atoms which are almost parallel to the electron beam*. This makes the geometry of electron diffraction patterns much simpler than that of X-ray diffraction patterns, for which θ can be very large.

3.1.2 The structure factor

Each type of plane in a crystal will have a different spacing and a different density of atoms per unit area, and so we may expect that the intensity of electron diffraction will be different for each type of plane. Although an accurate theoretical value can be obtained using the theory of dynamical electron diffraction, this is beyond the scope of this book. A simpler model, the theory of kinematical electron diffraction, can be used as a rough guide and correctly predicts those planes in a crystal which give zero diffracted intensity. The amplitude of an electron beam diffracted from the (hkl) planes of a unit cell in the crystal is

$$A_{hkl} \propto F_{hkl} A_0 \tag{3.5}$$

where A_{hkl} is the diffracted amplitude, A_0 is the amplitude of the incident beam, and F_{hkl} is the *structure factor* for the (hkl) planes. This structure factor is obtained simply by adding up the contribution to electron scattering made by each atom in the plane, taking into account the phase of each wave that is scattered. And since crystals are periodic, this only has to be done for one unit cell of the crystal, not a complete specimen. The structure factor is given by the equation

$$F_{hkl} = f_1(\theta) \exp[-2\pi i(hu_1 + kv_1 + lw_1)]$$
$$+ f_2(\theta) \exp[-2\pi i(hu_2 + kv_2 + lw_2)]$$
$$+ \ldots$$
$$+ f_n(\theta) \exp[-2\pi i(hu_n + kv_n + lw_n)]$$

where we count the atoms in the unit cell 1, 2, 3, ... n, and the jth atom has co-ordinates (expressed as a fraction of the axes of the unit cell) u_j, v_j, w_j. The factor $f_j(\theta)$ is simply the scattering factor for the jth atom, which gives the amplitude of scattering in the right direction from the atom, and the exponential factor keeps a track of the phase of each scattered wavelet. If there are n atoms in a unit cell, we obtain

$$F_{hkl} = \sum_{j=1}^{n} f_j(\theta) \exp[-2\pi i(hu_j + kv_j + lw_j)]. \tag{3.6}$$

In some cases, the individual terms can sum to zero, i.e. the planes give rise to no diffracted beam. These are known as *forbidden reflections*. As an example, consider the 100 reflection from a face-centred cubic (fcc) material such as gold. In this crystal structure, there are atoms at each corner of the cubic unit cell and in the middle of each face, and the group of four atoms at $[0, 0, 0]$, $[0, \frac{1}{2}, \frac{1}{2}]$,

Table 3.1 The first 12 allowed reflections in the fcc crystal structure and their interplanar spacings

Indices	$h^2 + k^2 + l^2$	d/a	Indices	$h^2 + k^2 + l^2$	d/a	Indices	$h^2 + k^2 + l^2$	d/a
111	3	$1/\sqrt{3}$	222	12	$1/3\sqrt{2}$	440	32	$1/4\sqrt{2}$
200	4	$1/2$	331	19	$1/\sqrt{19}$	531	35	$1/\sqrt{35}$
220	8	$1/2\sqrt{2}$	422	24	$1/2\sqrt{6}$	442	36	$1/6$
311	11	$1/\sqrt{11}$	333	27	$1/3\sqrt{3}$	533	43	$1/\sqrt{43}$

$[\frac{1}{2}\,0,\,\frac{1}{2}]$ and $[\frac{1}{2},\,\frac{1}{2},\,0]$ is the minimum set which describes the crystal completely. Using these four atoms in equation 3.6 gives

$$F_{hkl} = f(\theta) \begin{pmatrix} \exp[-2\pi i(1 \cdot 0 + 0 \cdot 0 + 0 \cdot 0)] \\ +\exp[-2\pi i(1 \cdot 0 + 0 \cdot \frac{1}{2} + 0 \cdot \frac{1}{2})] \\ +\exp[-2\pi i(1 \cdot \frac{1}{2} + 0 \cdot 0 + 0 \cdot \frac{1}{2})] \\ +\exp[-2\pi i(1 \cdot \frac{1}{2} + 0 \cdot \frac{1}{2} + 0 \cdot 0)] \end{pmatrix} = f(\theta) \begin{pmatrix} \exp[0] \\ +\exp[0] \\ +\exp[-\pi i] \\ +\exp[-\pi i] \end{pmatrix} = 0$$

$$(3.7)$$

The combination of the structure factor equation and Bragg's law allows the direction of all strongly diffracted beams from a crystal to be calculated. The structure factor equation is often summarized in diffraction rules – for example in the case of an fcc crystal, only those planes with all even or all odd Miller indices give diffracted beams (Table 3.1).

3.1.3 Diffraction in the transmission electron microscope

Electron diffraction patterns such as we are considering are normally obtained with a transmission electron microscope (TEM), an instrument which we will be considering in more detail in Chapter 4. The TEM can provide two separate kinds of information about a specimen – a magnified image, and a diffraction pattern. The way in which this comes about can be seen by examining the electron optics of a simple projection microscope.

Figure 3.3 shows two sets of parallel rays leaving the specimen AA′ in a two-stage projection microscope. Before forming the first image at BB′, all rays parallel to the two single-arrowed rays will pass through the point D′ and all rays parallel to the double arrowed rays will pass through D. For every set of parallel rays leaving the specimen there is a corresponding point in the plane DD′ (known as the back focal plane of the objective lens). Similarly, if we follow the ray paths through the second (projector) lens we find that a second set of points is formed in the plane EE′. If we were to put a viewing screen at DD′ or EE′ instead of at CC′ we would see the *diffraction pattern* instead of the image.

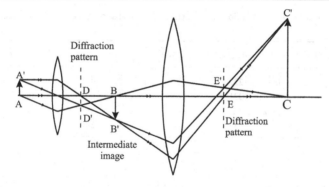

Figure 3.3 The ray diagram of a two-stage projection microscope showing the positions of the diffraction pattern (DD' and EE') and image (BB' and CC').

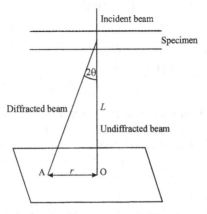

Figure 3.4 Schematic diagram showing the geometry of diffraction pattern formation.

In order to understand the geometry of electron diffraction, we can ignore the lens system, which merely magnifies the diffraction pattern, and consider the much simpler ray diagram of Figure 3.4.

A beam of electrons impinges on a crystalline specimen. Some of the electrons pass through the specimen without interaction, and hit the screen which is at a distance L from the specimen, at O. Other electrons are diffracted through an angle 2θ by the crystal planes of spacing d, and these electrons hit the film at A, which is a distance r from O. From simple geometry, we see that for small angles of diffraction

$$\frac{r}{L} = 2\theta \tag{3.8}$$

Combining this with equation 3.4, we find

$$\frac{r}{L} = \frac{\lambda}{d}, \quad \text{or} \quad rd = L\lambda \tag{3.9}$$

As the camera length L and the electron wavelength λ are independent of the specimen, and are a constant for the instrument, we call $L\lambda$ the *camera constant*. It can be seen that the distance of a diffraction spot from the undiffracted spot, r, is therefore inversely proportional to the d spacing of the diffracting planes. If we know the camera constant for the instrument, then we can determine d simply by measuring r on the pattern. In a real microscope, because of the lenses between the specimen and the screen, L is not the physical distance between the specimen and the screen, but is a notional distance which can be changed by the microscopist.

3.2 Diffraction spot patterns

In section 3.1.1 we showed that the planes of a crystal would diffract electrons if these planes were lying approximately parallel to the electron beam. Thus, as shown schematically in Figure 3.5(a), a single crystal specimen oriented such that several sets of planes are parallel to the beam will give rise to a diffraction pattern consisting of a regular array of spots. If the specimen contains several crystals of different orientations, as in Figure 3.5(b), then the diffraction pattern is the sum of the individual patterns, and is more complicated. As shown in section 3.1.2 only certain crystal planes can diffract; this means that the number of possible d spacings and hence r spacings on the pattern is limited, and the spots are not randomly distributed but fall on rings (each of which has constant r).

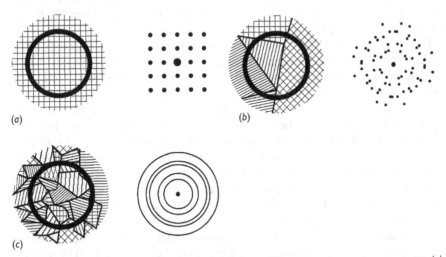

Figure 3.5 Types of diffraction pattern which arise from different specimen microstructures. (a) A single perfect crystal. (b) A small number of grains – notice that even with five grains the spots begin to form circles. (c) A large number of randomly oriented grains – the spots have now merged into rings.

Figure 3.5(c) shows the case for a specimen containing very many crystals of random orientation. In this situation, the spots on the rings are so close together that the rings appear continuous, and the diffraction pattern is of the type shown in Figure 3.1(c).

In the discussion above, we have assumed that the diffraction pattern is obtained from the whole specimen, but in practice this is not the case. Either by using suitable apertures, or by illuminating only a small part of the specimen, it is possible to obtain a diffraction pattern from a very small region – such as just one grain in a polycrystalline sample, or a precipitate of one material lying in a matrix of another material. The way in which this is carried out in the transmission electron microscope is discussed in Chapter 4.

In the following sections, the principles involved in solving diffraction patterns, and some examples of the type of information which is obtainable, are discussed. There are many textbooks and handbooks available which will provide much more detailed information for those who wish to pursue the subject further.

3.2.1 Analysing diffraction patterns

There are many reasons for wanting to analyse a diffraction pattern. It may be necessary to accurately measure the camera constant of the microscope, or to attempt to identify an unknown material using its interplanar spacings or diffraction geometry. Or a certain diffraction condition may be needed to obtain a particular diffraction contrast image (see Chapter 4), or the exact orientation of the crystal needed to find out the habit plane of a particular type of defect. The accuracy and type of measurement made depend both on the specimen and the information required.

The accuracy of a TEM lattice spacing measurement is not as good as from modern X-ray techniques, because it is harder to measure the angle of diffraction to the same precision and very difficult to produce identical imaging conditions from sample to sample. The most accurate measurements are best made in conjunction with a standard of known lattice parameter. Both sample and standard must be accurately at eucentric height (see section 4.1) when the two diffraction patterns are recorded, and no adjustments to diffraction focus or camera length made between the two exposures. Several computer packages are now available which can measure the distance between spots or ring radii more accurately than can be achieved with a ruler (about 0.1%). Even without taking such elaborate precautions, it is possible to obtain interplanar spacings accurate to one or two percent, which is accurate enough for most purposes – such as comparing measured values with those of various candidate materials held in the JCPDS powder diffraction index of known crystalline materials. In what follows, we give examples of the most common types of measurement and describe typical procedures used to obtain the different kinds of information.

Camera constant calibration is most often performed using a polycrystalline gold foil; such samples are easily prepared by evaporation of gold onto a very thin carbon or formvar film, and have a grain size small enough to give a continuous ring pattern such as Figure 3.1(c). Gold has the fcc crystal structure, with lattice parameter $a = 0.4078$ nm. The rings can be indexed according to Table 3.1. Since the radius of the ring r is inversely proportional to the plane spacing d, the innermost ring, 111, has the largest plane spacing (0.2354 nm). Since d is known for each ring, a measurement of r gives the camera constant $L\lambda$ directly from equation 3.9. Checking that all the rings give the same value for the lattice parameter is a good way to check that the diffraction pattern is accurately in focus.

A polycrystalline ring pattern from an unknown material is essentially similar to an X-ray Debye–Scherrer pattern, and is solved in a similar manner. It can provide information about the crystal structure and lattice spacing of the sample. As described in Section 3.1.2, only certain planes in a crystal will diffract electrons. It is relatively straightforward in the case of cubic materials to obtain the crystal type (i.e. primitive, body centred, or face centred) from the diffraction rules that apply (e.g. Table 3.1), and hence to index the rings simply by inspection. In most cases, information on the composition of the material can be used to reduce the problem from identifying a completely unknown compound to choosing between a few known ones. The composition of the material is often determined in the TEM using energy dispersive X-ray spectroscopy, as described in Chapter 6. The combination of diffraction and compositional information available in a TEM makes it a very powerful tool for the identification of unknown materials.

Single crystal diffraction patterns are often used to determine the crystallographic orientation of the specimen in the microscope. This can be important when analysing crystallographic defects in the material, since it is nearly always necessary to obtain a particular diffraction condition (see Chapter 4) to make the defects visible and to understand their nature.

Starting with the pattern of Figure 3.1(d), from aluminium ($a = 0.405$ nm), and with a camera constant of 1.46×10^{-12} m², one procedure is as follows. The table below shows the results obtained at each stage of the analysis:

	(a) Measured r values	(b) Measured angle (α)	(c) Calculated d spacings	(d) General indices	(e) Particular indices
Spot 1	6·25 mm	90°	0·234 nm	111	111
Spot 2	10·20 mm		0·143 nm	220	$\bar{2}$20

(a) The r values for two diffraction spots close to the centre of the diffraction pattern are measured.

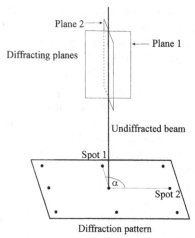

Figure 3.6 Schematic diagram of a diffraction pattern from a single crystal.

(b) The angle α which the two spots subtend at the undiffracted spot is measured (Figure 3.6).
(c) If the camera constant is known, then d_1 and d_2 are determined.
(d) The indices of the diffraction spots are determined from equation 3.5. We have only determined the general nature of the indices, e.g. spot 2 in Figure 3.6 is 220 type. We have not yet determined the particular indices, i.e. whether this spot is actually 220, $2\bar{2}0$, $\bar{2}20$, $\bar{2}\bar{2}0$, 202, $\bar{2}02$, $20\bar{2}$, $\bar{2}0\bar{2}$, 022, $02\bar{2}$, $0\bar{2}2$ or $0\bar{2}\bar{2}$.
(e) We have freedom to choose the particular indices of one spot. For example, we can take spot 1 to be 111. Since, in this case, the angle α is 90° and the crystal system is cubic, we know that the vector dot product of the indices of the spots must be zero, and the indices of spot 2 must thus add up to zero. This restricts the indices of spot 2 to be one of $2\bar{2}0$, $\bar{2}20$, $\bar{2}02$, $20\bar{2}$, $02\bar{2}$, or $0\bar{2}2$. Once we choose one of these directions, we have fixed the indexing of all subsequent diffraction patterns and crystallographic directions on images.

In practice, there are a few trivial factors which can make consistent indexing of more than one diffraction pattern, or matching a diffraction pattern to an image, a confusing process. The first is the rotation of a diffraction pattern relative to its corresponding image. As described in section 2.4, the image produced by a magnetic lens is rotated relative to the sample. Since the lens currents in the microscope must differ when obtaining an image or a diffraction pattern, there will be a relative rotation. Furthermore, the image rotation will depend upon the magnification used, and the diffraction pattern rotation will depend upon the camera length used. Each microscope will usually have the relative rotations between image and diffraction pattern tabulated somewhere, but it is sound advice to pick a camera length and image magnification and stick to them whenever a self-consistent set of images and patterns is needed. A second source of confusion is the difference between the recorded image and the view on the microscope screen. Film is placed in the microscope camera with its emulsion side up, facing the electron beam. It is then usually printed or scanned with its emulsion side down, towards the photographic paper or CCD array. This produces a print which is a mirror image of what the microscopist sees on the screen, and can make it hard to relate the notes made while on the microscope to the images in front of you. The important thing is to be aware of these simple pitfalls before you start, so that you can take the simple measures needed to deal with them.

3.3 Use of the reciprocal lattice in diffraction analysis

Although, as we have seen in section 3.2, we can understand the principle of diffraction patterns from simple geometry, if we wish to go deeper into the subject, and in particular to understand the intensity of electron diffraction, it is very convenient to discuss diffraction in terms of the *reciprocal lattice*. The concept of the reciprocal lattice is quite straightforward and is used widely in crystallography, diffraction and solid state physics, and comprehensive treatments can be found in textbooks on these subjects.

Figure 3.7(a) shows part of a crystal lattice, a set of planes lying perpendicular to the paper. In the reciprocal lattice, each plane of the real lattice is represented by a point, lying a distance $1/d$ from the origin (O), on the perpendicular to the plane. The vector describing the position of the reciprocal lattice point relative to O is often denoted by **g**, commonly known as the *diffraction vector*. Therefore for the real lattice of Figure 3.7(a), the corresponding reciprocal lattice is as shown in Figure 3.7(b). For a three-dimensional lattice, we can clearly construct a corresponding three-dimensional reciprocal lattice.

If we apply the same transformation to the reciprocal lattice, the real lattice is obtained once more. The equivalence of the two representations of the lattice

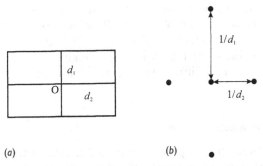

(a)   (b)

Figure 3.7 A two-dimensional representation of (a) a real lattice, and (b) the corresponding reciprocal lattice.

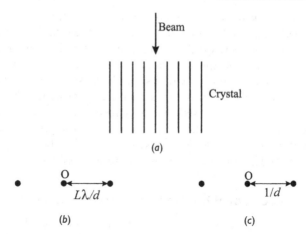

Figure 3.8 The relationship between electron diffraction and the reciprocal lattice. (a) The crystal, (b) the diffraction pattern, (c) the reciprocal lattice.

in Figure 3.7 is shown by the fact that if we know either the real or reciprocal lattice, we can construct the other.

In order to see why the reciprocal lattice should be of use in diffraction, let us first recall what we have learnt about electron diffraction.

Diffraction occurs from planes which are approximately parallel to the electron beam (Figure 3.8(a)).

The diffraction pattern consists of points, spaced a distance proportional to $1/d$, aligned in the direction perpendicular to the planes (Figure 3.8(b)).

If we consider the lattice planes of Figure 3.8(a), then we can construct the reciprocal lattice – Figure 3.8(c), and see that it is a row of points of spacing $1/d$, lying perpendicular to the planes. Thus we see that there is a strong resemblance between the diffraction pattern and the reciprocal lattice. In fact the diffraction pattern is, to a first approximation, a scaled section through the reciprocal lattice normal to the beam. From equation 3.6, we see that the scaling factor is $L\lambda$, the camera constant.

We therefore now have a powerful way of constructing the electron diffraction pattern for any known crystal. We simply have to construct the reciprocal lattice of the crystal, and we can obtain the diffraction pattern for any orientation of the crystal by orienting the lattice accordingly, and taking a section through it (remembering that diffraction rules will cause certain planes not to diffract).

3.3.1 The Ewald sphere construction

The relationship between the reciprocal lattice and the diffraction pattern can be demonstrated rather more formally by the *Ewald sphere construction*, which is shown in Figure 3.9.

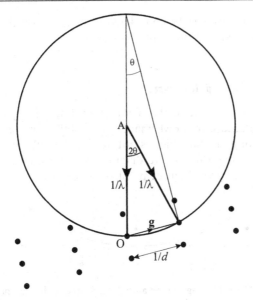

Figure 3.9 The Ewald sphere construction.

(a) The diffracting crystal is represented by its reciprocal lattice.
(b) The electron beam is represented by a vector of length $1/\lambda$, parallel to the beam direction, and terminating at the origin of the reciprocal lattice.
(c) A sphere (the Ewald, or reflecting sphere) of radius $1/\lambda$ is drawn about A.

As Figure 3.9 is drawn, the Ewald sphere passes through a reciprocal lattice point a distance $1/d$ from the origin. From geometry, we find that

$$\sin \theta = \frac{(1/d)/2}{1/\lambda} = \frac{\lambda}{2d}$$

or

$$\lambda = 2d \sin \theta$$

i.e. Bragg's Law is satisfied by this construction.

Thus we see that we can restate the conditions for diffraction as: *diffraction occurs when the Ewald sphere touches a reciprocal lattice point.*

Although this is not quite the same definition of diffraction as that which we deduced in the last section, the two statements are in practice equivalent for the diffraction of electrons by crystals, because the radius of the Ewald sphere is large ($270\,\text{nm}^{-1}$ for $100\,\text{keV}$ electrons) compared to reciprocal lattice vectors (typically about $5\,\text{nm}^{-1}$). Therefore, to a first approximation, for small angles of diffraction, the Ewald sphere can be considered to be a plane. Thus, we see why a diffraction pattern can be considered to be a section through the reciprocal lattice, as we deduced previously.

3.3.2 Diffraction from a finite crystal

In order to understand some further features of diffraction patterns, and also to understand the nature of TEM images of crystalline materials, we need to consider the case of a crystal which is not oriented exactly at the Bragg angle. Figure 3.10 shows an Ewald sphere diagram for this situation, the reflecting sphere missing the reciprocal lattice point by a vector s.

Figure 3.10 The Ewald sphere diagram for a crystal oriented such that the reciprocal lattice misses the reflecting sphere by a vector s.

On the basis of our considerations so far, we would predict that there would be no diffraction under these conditions because we have shown that diffraction only occurs when the reflecting sphere touches the reciprocal lattice. However, for real crystals, this condition can be relaxed, resulting in significant diffracted intensity.

If we make the assumption that diffraction is weak, i.e. that the probability of any electron being diffracted is small, then the intensity of diffraction from a crystal of finite thickness may be calculated using the *kinematical theory of electron diffraction*.

Consider an electron beam penetrating a specimen of thickness t as shown in Figure 3.11. The crystal is notionally divided into slices perpendicular to the electron beam, and the amplitude and phase of the electron scattering are calculated for each slice. The intensity of scattering from the crystal (I_g) is then obtained by summing the scattering from each slice, taking into account phase differences of the waves scattered at different depths. It can be shown that the diffracted intensity is given by:

$$I_g = \left(\frac{\pi}{\xi_g}\right)^2 \frac{\sin^2(\pi t s)}{(\pi s)^2}.$$ (3.10)

ξ_g is a constant of the material for a particular value of $\mathbf{g}$, and is known as the extinction distance. It is given by:

$$\xi_g = \frac{\pi V_c \cos \theta_B}{\lambda F_g}.$$ (3.11)

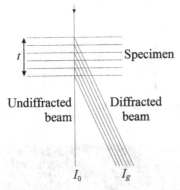

Figure 3.11 Schematic diagram showing the path differences between electrons scattered at different depths in a crystal.

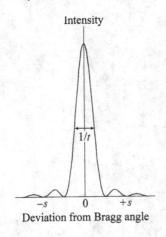

Figure 3.12 The variation of diffracted intensity with deviation from the Bragg angle for a crystal of thickness *t*.

where V_c is the volume of the unit cell, λ is the electron wavelength, θ_B is the Bragg angle and F_g is the structure factor. The extinction distance is a crucially important parameter which determines many of the characteristics of images which rely on diffraction of electrons into or out of the image. For example, the amount of contrast at a particular thickness of the specimen, the apparent size of a defect and the appearance of stacking fault fringes are all determined by ξ_g.

Figure 3.12 shows a graphic representation of the variation of diffracted intensity I with the deviation, s, from the Bragg angle, θ_B, for a specimen of constant thickness.

We see that this theory predicts that although the intensity of diffraction is a maximum at the Bragg angle ($s = 0$), there will be some diffraction when s is non-zero. The width of the main peak at half height is, as indicated on the

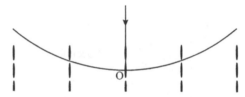

Figure 3.13 The Ewald sphere construction for a thin crystal.

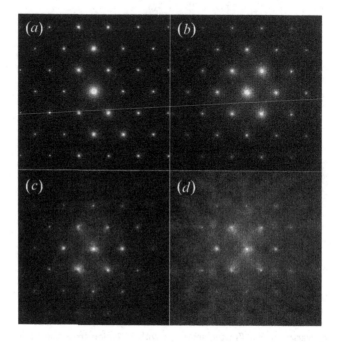

Figure 3.14 Electron diffraction patterns from a silicon crystal which is oriented such that the incident beam is parallel to a prominent zone axis [110]. The patterns (a) to (d) are taken from progressively thicker regions of crystal.

diagram, equal to $1/t$. Therefore we see that the thinner the crystal, the further the crystal may deviate from the Bragg condition and yet diffract.

In terms of the Ewald sphere construction, we can represent this relaxation of the diffraction conditions for a thin crystal as shown in Figure 3.13 in which the reciprocal lattice points are now extended in a direction *normal to the plane of the specimen*. Comparing Figures 3.10 and 3.13, we find that the chances of the Ewald sphere touching the rods of the reciprocal lattice (i.e. of diffraction occurring) are now increased. Since the length of the rods is inversely proportional to the specimen thickness, we see that the thinner the specimen, the more diffraction spots will occur in the pattern. Figure 3.14 is a series of diffraction

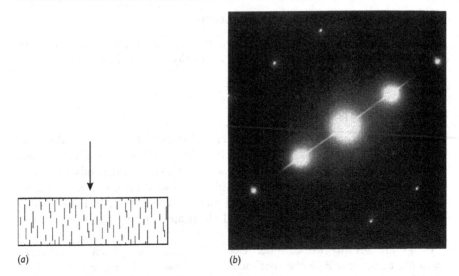

(a)　　　　　　　　　　(b)

Figure 3.15 Diffraction from a specimen containing thin crystalline plates which are parallel to the electron beam. (a) Schematic representation of the specimen. (b) Streaked diffraction pattern from an aluminium–copper alloy, heat treated to produce coherent plate-shaped particles (GP zones), which are oriented as shown in (a).

patterns taken from the same wedge-shaped crystal, at different thicknesses, and clearly demonstrates the reduction in the number of diffraction spots as the specimen thickness increases.

In a thin crystal, electron diffraction can occur even if the specimen is tilted up to five degrees from the Bragg angle. This means that the orientation of a crystal can only be determined with an accuracy of a few degrees from spot patterns. Fortunately, there are other diffraction effects which allow much more accuracy, i.e. convergent beam electron diffraction patterns and Kikuchi lines (see section 3.4).

The idea of representing variations in electron diffraction conditions by placing a rod at each reciprocal lattice point can be generalized to deal with a crystal that is thin in any dimension. In this case, the diffraction condition at a reciprocal lattice point is extended in each dimension by an amount inversely proportional to the thickness of the crystal plane in that dimension.

For example, for a plate-like crystal oriented parallel to the beam (Figure 3.15(a)), the reciprocal lattice points are extended perpendicular to the beam, and the diffraction pattern – which is the intersection of the Ewald sphere and the reciprocal lattice – will show streaking of the spots. An example of a diffraction pattern from a crystal of aluminium containing thin plate-like particles of a second phase oriented parallel to the beam is shown in Figure 3.15(b).

3.4 Other types of diffraction pattern

Although spot diffraction patterns of the type discussed above are often the most commonly used, there are other diffraction effects which yield useful information.

3.4.1 Kikuchi line patterns

As may be seen from Figure 3.14, as the crystal thickness increases, so does the diffuse background of the pattern, which is due to the inelastically scattered electrons. The intensity of inelastically scattered electrons depends upon the angle of scattering, and is a maximum in the forward direction as shown in Figure 3.16. In a crystalline specimen, some of these inelastically scattered electrons may subsequently be scattered elastically, and it is this interaction which gives rise to Kikuchi lines.

In Figure 3.17(a) we consider electrons which have been inelastically scattered at a point P in the specimen. Now some of these electrons will be travelling at the Bragg angle (θ) to planes Q and R, and may therefore undergo Bragg diffraction. Thus the inelastically scattered ray of intensity I_S – which would have gone to S – is now diffracted to T, and the inelastically scattered ray of intensity I_T which would have gone to T is now diffracted to S. Thus the intensity of inelastic scattering is modified from that predicted by Figure 3.16.

At point S, the intensity is increased by $I_T - I_S$, and at point T, the intensity is increased by $I_S - I_T$. As OS is less than OT, it follows from Figure 3.16 that I_S is larger than I_T. Therefore there is a net decrease in intensity of inelastic scattering at S, and a net increase at T, as shown in Figure 3.17(b). In the two-dimensional diffraction pattern this effect appears as a pair of parallel lines, a dark one ('defect' line) at S, and a bright one ('excess' line) at T. Figure 3.18(a) shows an example of Kikuchi lines in an aluminium specimen.

The angular separation of the lines is seen from Figure 3.17 to be 2θ, which is of course the same as the separation of the diffraction spots from the same planes. Diffraction spots and Kikuchi lines often occur in the same pattern, although the intensity of Kikuchi lines increases as the specimen thickness

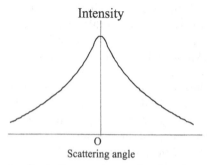

Figure 3.16 The intensity of inelastic scattering as a function of scattering angle.

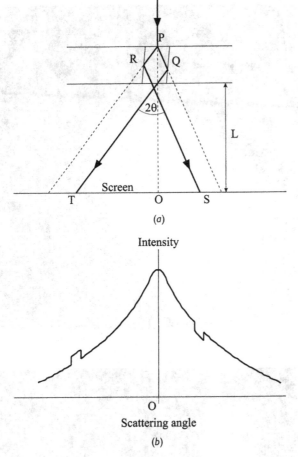

Figure 3.17 (a) Ray diagram showing the geometry of Kikuchi lines (see text for details). (b) The intensity of inelastic scattering when Kikuchi lines are produced.

increases, and, as discussed above, the intensity of the diffraction spots decreases as the specimen thickness increases. Figure 3.18(b) shows an example of Kikuchi lines and diffraction spots.

It may be seen from Figure 3.17 that the Kikuchi lines are symmetrically astride the diffracting planes, and if the specimen is tilted, then the lines will move as if they are rigidly fixed to the specimen. For a small tilt α, the lines will move a distance δ across the diffraction pattern, and from simple geometry it can be seen that $\delta = L\alpha$. For a typical camera length of 100 cm, this means a shift of 30 mm per degree of tilt. The position of Kikuchi lines may be used to determine orientations to within a small fraction of a degree, and may also be used to measure small orientation changes within a specimen, such as across a low angle grain boundary. Figures 3.18(b) and (c) show the effect of a small tilt

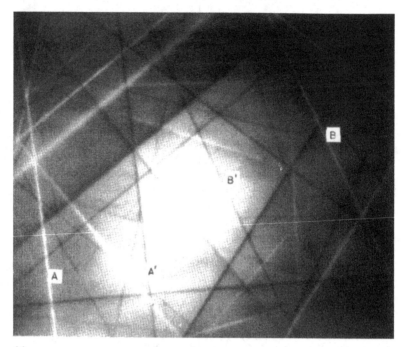

(a)

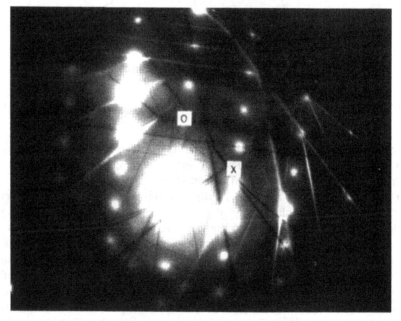

(b)

(c)

Figure 3.18 (*a*) Kikuchi lines in an aluminium specimen. Note the pairs of parallel light and dark lines such as AA′ and BB′. (*b*) Diffraction pattern showing Kikuchi lines and diffraction spots in an aluminium specimen. (*c*) Diffraction pattern of the same sample after tilting the specimen by approx. 1°. By identifying a particular diffraction spot, such as O, and a particular intersection of Kikuchi lines (e.g. X), it can be seen that this small tilt has caused a considerable shift of the Kikuchi line pattern relative to the spot pattern.

on the diffraction pattern. Note that although the Kikuchi lines move across the pattern during tilting, the position of the diffraction spots remains fixed.

3.4.2 Convergent beam electron diffraction (CBED) patterns

The spot diffraction patterns discussed earlier in this chapter are normally obtained with a parallel beam of electrons illuminating the specimen. However, if a convergent electron beam is focused on the specimen, then, as shown in Figure 3.19, the diffraction spots become discs, and from an analysis of their fine structure it is possible to obtain information about the specimen thickness, crystal structure and lattice parameters.

The quality of the pattern is best if the diffracting region is of uniform thickness and orientation and is free of defects. Therefore only a small portion of the specimen, of a diameter no more than a few tens of nm, is generally illuminated.

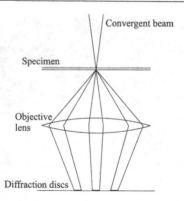

Figure 3.19 Ray diagram for convergent beam diffraction.

The use of such a small probe size makes the technique very suitable for the study of small particles, and, as a crystallographic tool, it complements X-ray methods which can obtain similar information, but only from large samples.

In section 3.3.1 we interpreted the diffraction pattern as being a section through the reciprocal lattice, and showed that we could ignore the curvature of the Ewald sphere. However, for relatively large angles of diffraction this is not the case. In Figure 3.20 we see that the sphere will intersect other layers of the reciprocal lattice and, if the collection angle is sufficiently large, this will result in corresponding rings of diffraction spots, known as *higher order Laue zone (or* HOLZ) rings.

The importance of these zones is that they provide information about the crystal in the direction parallel to the electron beam, in addition to the two-dimensional information obtainable from the zero order Laue zone, which is the one used for normal spot patterns.

An example of a CBED pattern showing several HOLZ rings is shown in Figure 3.21.

When higher order Laue zones are excited, fine lines due to diffraction from the higher order Laue zones are generally seen within the diffraction discs as shown in Figure 3.22. These HOLZ lines are formed in a very similar manner to Kikuchi lines (Figure 3.17), and their position is very sensitive to the crystal lattice parameter. They can be used to measure relative or absolute lattice parameters, the latter to an accuracy of 1 part in 10^4, and hence determine indirectly local lattice strains or changes in chemical composition. However caution must always be exercised since the lattice parameters of a thin specimen may be slightly different from those of a bulk specimen, especially if there is internal strain in the specimen.

From the overall symmetry of CBED patterns obtained by aligning the incident electron beam along several prominent crystallographic directions – often referred to as *zone axes* – it is possible to obtain the point and space

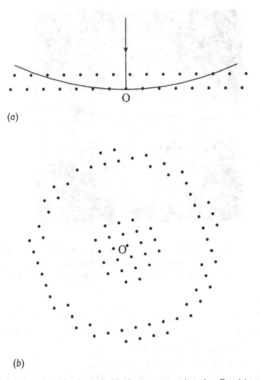

(a)

(b)

Figure 3.20 (a) Intersection of higher order Laue zones by the Ewald sphere. (b) Schematic representation of the resulting diffraction pattern.

Figure 3.21 Convergent beam diffraction pattern from a [111] oriented gamma prime particle in a nickel-base superalloy. The pattern shows the diffraction discs of the zero order Laue zone, and also excitation of higher order Laue zones (HOLZ rings). (Yang-pi-Lin, University of Bristol.)

Figure 3.22 Part of a convergent beam diffraction pattern showing the [1 1 3] bright field disc of silicon. The dark lines crossing the disc are HOLZ lines, whose positions are very sensitive to the lattice parameter of the specimen (Yang-pi-Lin, University of Bristol).

groups of the crystal. For this type of work it is often useful to compare the patterns obtained with those from known structures, and compilations of CBED symmetries for this purpose are available.

CBED patterns can also be used to measure important parameters such as the operating voltage of the microscope and the specimen thickness. The latter is described further in section 4.2.

3.5 Questions

1 Draw (a) The Ewald sphere construction, and (b) the undeflected and diffracted electron beams and the operating planes, when planes characterized by the reciprocal lattice vectors **g**, **2g** and **3g** lie in the Bragg condition.

2 What is the Bragg angle for first order and second order diffraction from planes with $d = 0.08934$ nm and electron wavelength 0.00164 nm (accelerating voltage $= 400$ kV).

3 What do you expect to happen to an electron diffraction pattern as the accelerating voltage is increased?

4 What are the five lowest-index diffracting planes for (a) a primitive cubic crystal? (atom position $= [0\ 0\ 0]$), and (b) a body centred cubic crystal? (atom positions $= [0\ 0\ 0]$, $[0.5\ 0.5\ 0.5]$].

5 The sum of the indices of the spots in a diffraction pattern taken with the incident electron beam parallel to [111] direction in an fcc crystal must add up to zero. (a) What are the indices of the spots closest to the undeflected beam? (b) If the distance between the spots is 12 mm and the camera constant is 1.8×10^{-12} m^2, what is the lattice parameter of the material?

6 How would you obtain an electron diffraction pattern from one grain in a polycrystalline specimen if the average grain size is (a) 5 μm (b) 20 nm?

7 Your microscope has a severe contamination problem which results in the deposition of an amorphous organic film on the specimen in areas illuminated by the electron beam. How would this manifest itself in a diffraction pattern over time?

8 If you can see a shift of 0·1 mm in the position of a Kikuchi line on your microscope screen, how accurately can you align a specimen to a given crystallographic direction using a camera length of 46 cm?

9 How do the HOLZ rings of a diffraction pattern obtained using selected area diffraction differ from those obtained using convergent beam diffraction?

10 What is the radius of the first HOLZ ring for a material with planar spacing (parallel to the beam) 0·0600 nm and a camera constant of $3·0 \times 10^{-12}$ m^2?

Chapter 4

The transmission electron microscope

4.1 The instrument

In Chapter 2 most of the essential parts of a TEM were described. The illumination is provided by an electron gun while the lenses are all electro-magnetic and work as described in section 2.4. The remaining necessary components are a viewing screen – usually a simple layer of electron-fluorescent material, viewed through a lead glass window – and a camera, which must work in the vacuum within the microscope. These components are assembled into a vertical 'microscope column' of which a typical example is shown in Figure 4.1. Most of the thickness of the column is taken up with the windings of the lens coils, the pole pieces, and pipes of cooling water – the electrons

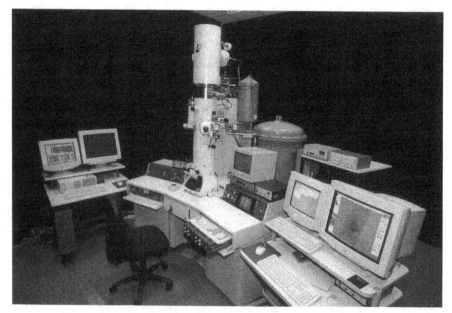

Figure 4.1 A modern transmission electron microscope. (Courtesy of Professor A. G. Cullis)

travel down a fine tube in the centre which rarely exceeds a millimetre or so in diameter.

Early microscopes generally used one *condenser* lens as a collimator to control the electron beam before it reached the specimen, and one or two *objective* and *projector* lenses to magnify the subsequent image. They were therefore exact electron-optical analogues of Figure 1.4(*a*). Nowadays it is more common to find two condenser lenses and four or five projector lenses. The electronics needed to control the electron gun, six or seven lenses and alignment coils is quite complex and the microscope shown in Figure 4.1 is a mass of knobs and dials. However, once the optical principles are understood even the most sophisticated TEM is very simple to operate. There is a tendency for modern microscopes to be controlled by a built-in computer and this results in a reduction in the number of visible controls. However it does not affect the optics of the microscope or the number of parameters which must, explicitly or implicitly, be controlled.

We will now examine a modern microscope in more detail, using the instrument shown in Figure 4.1 as an example.

4.1.1 The electron gun

At the top of the instrument, about a metre above the seated operator's head, is the electron gun. The most common types of TEM have thermionic guns (section 2.3) capable of accelerating the electrons through a selected potential difference in the range 40–200 kV. The appropriate electron energy depends upon the nature of the specimen and the information required. For some applications, particularly if the specimen is relatively thick or very high resolution is required, it is an advantage to use much higher electron energies. A range of *medium voltage* (300–400 kV) and *high voltage* (600–3000 kV) microscopes have been developed for these purposes. The very high energy microscopes have become rarer as the resolution of lower energy microscopes has improved through better lens design and as specimen preparation techniques have developed. Furthermore, field emission guns have now developed to the point that very fine electron beams can be produced (of the order of 1 nm at the specimen) and they are becoming increasingly widespread.

4.1.2 The condenser system

Below the electron gun are two or more condenser lenses. Together, they demagnify the beam emitted by the gun and control its diameter as it hits the specimen (Figure 4.2). This enables the operator to control the area of the specimen which is hit by the beam and thus the intensity of illumination. An aperture is present between the condenser lenses (the *condenser aperture*) which can be used to control the convergence angle. At their simplest the condenser controls can be thought of as brightness controls but in fact they

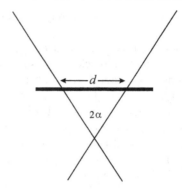

Figure 4.2 A beam coming to focus just after the specimen ('underfocused') illuminates an area of diameter *d*. The beam convergence is α.

permit a wide range of control over the region of specimen which is 'sampled' by the beam and over the type of diffraction pattern which is formed (section 3.4). It is useful to consider the widely used two-condenser illumination system in more detail. The first condenser lens (C1), often labelled *spot size*, sets the demagnification of the gun crossover (section 2.3). The second lens (C2, often *intensity*) provides control of the convergence angle of the beam leaving the condenser assembly, as Figure 4.3 shows. A parallel beam will illuminate a large area of the specimen (rays (i)); as the convergence angle is increased, the beam diameter at the specimen decreases until it reaches its minimum (rays (ii)), when the beam is focused on the specimen. As the convergence angle is increased still further, the beam focus is before the specimen and the illuminated area increases once more (rays (iii)).

The minimum possible illuminated area is controlled by the effective size of the source at the gun, the strength of C1 and the condenser aperture. It cannot of course be made any smaller than the limit imposed by diffraction at the condenser aperture and by aberrations in the condenser system.

When the beam is focused on the specimen, its convergence angle is controlled by the condenser aperture. The size of the condenser aperture also affects image quality, since the electrons which pass far from the optic axis will be strongly affected by spherical aberration (see Chapter 1). It also has a large effect on the intensity of the electron beam. The illumination provided by the condenser lens must be varied by the microscopist according to the kind of information he is trying to capture. As in most things, a balance must be sought to get the best result. Thus a diffraction contrast image is usually taken with a medium size condenser aperture, and fairly large spot size to maximize illumination (but not so large as to degrade image quality), and a close to parallel beam, to give even illumination and similar diffraction conditions (but sufficiently intense to avoid very long exposure times, which could cause problems due to specimen drift). Alternatively, a convergent beam

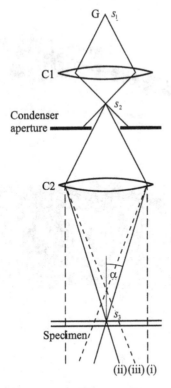

Figure 4.3 The two-lens condenser system. The spot of size s_1 at the gun crossover (G) is de-magnified to s_2 by the first condenser lens C1. The second condenser lens C2 is used to focus the beam; it may also change the spot size to s_3. (i), (ii) and (iii) show un-derfocused, focused and overfocused beams respectively. The convergence angle α is controlled by the condenser aperture. An animated version of this diagram can be found in MATTER: Introduction to electron microscopes.

electron diffraction (CBED) pattern would need a small spot size (to reduce the effects of bending and defects in the sample), a large condenser aperture (to give a large disk in the CBED pattern) and the beam focused accurately onto the specimen using C2.

The MATTER software module 'The TEM' includes a model of a two-condenser system in which the strength of each lens can be varied independently and the effect on spot size and convergence can be observed.

4.1.3 The specimen chamber

Below the condenser lies the specimen chamber, which is one of the most crucial parts of the microscope. A very small specimen must be held in precisely the correct position inside the objective lens, but should also be

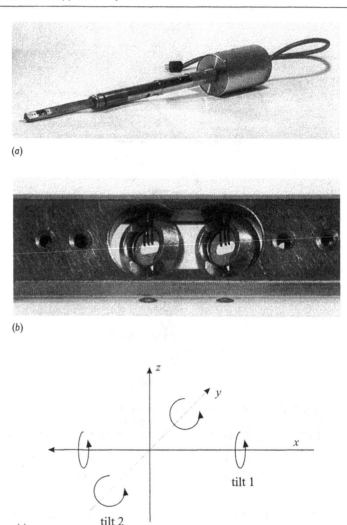

(a)

(b)

(c)

Figure 4.4 (a) A side entry double tilt holder. (b) A higher magnification view of the specimen cups. (c) The range of movements and tilts available with such a specimen holder. An animated diagram showing the effect of specimen tilt when the sample does not lie at the eucentric height is shown in MATTER: Introduction to electron microscopes.

capable of being moved several millimetres and tilted by large angles. Additionally, X-rays must be permitted to leave the area if the microscope is to be used in an analytical mode. These constraints are usually met by a side-entry specimen rod (e.g. Figure 4.4) which holds a 3 mm diameter specimen (or a smaller specimen on a 3 mm support grid) between the pole pieces of the objective lens. The specimen rod enters the column through an airlock, and can usually be moved up to 2 mm in the x and y directions to locate the region of interest, and by a fraction of a millimetre in the z direction, in order to bring it to the object plane of the lens. It is quite easy to tilt the specimen about the long axis of the holder by up to 60 degrees by rotating the holder itself. Tilt about an axis perpendicular to this is also very desirable but is achieved with more difficulty. The mechanisms which provide all these movements and tilts must ensure that when the appropriate specimen position has been selected it does not move by more than about 0·1 nm or less while the micrograph is exposed (depending upon the magnification and type of image). A movement of 0·1 nm in 1 second is equivalent to 1 mm in four months – the specimen must clearly be extremely stable.

An important consideration is that the specimen does not move laterally when it is tilted. This can only be true if the axis of tilt accurately intercepts the optical axis of the microscope, along which the beam is travelling. It is possible to adjust most specimen holders so that one tilt is in this *eucentric* position, but normally it is not possible to also make the second axis of tilt eucentric. This means that it may be necessary to use considerable skill to set up some of the diffraction conditions described in Chapter 3 and later in this chapter.

4.1.4 The objective and intermediate lenses

The objective lens is so strong that the specimen sits within its pole pieces. The role of the objective lens is to form the first intermediate image and diffraction pattern, one or other of which is enlarged by the subsequent projector lenses and displayed on the viewing screen.

The optics of the objective lens are shown in Figure 4.5. As has been emphasized in Chapter 3, a diffraction pattern is inevitably formed in the back focal plane of the lens. The first projector lens (often called the intermediate or diffraction lens) can usually be switched between two settings, shown in Figure 4.5(a) and (b). In the *image* mode, it is focused on the image plane of the objective (Figure 4.5(a)). The magnification of the final image on the microscope screen is then controlled by the strength of the remaining projector lenses (not shown in Figure 4.5). In the *diffraction* mode the intermediate lens is focused on the back focal plane of the objective (Figure 4.5(b)) and the diffraction pattern is projected onto the viewing screen. The magnification of the diffraction pattern is controlled by the projector lenses and is usually described in terms of the effective *camera length* of the system.

An essential feature of the objective system is the aperture holder which enables any one of three or four small apertures to be inserted into the column in the back focal plane. The objective aperture clearly defines the angular range of scattered electrons which can travel further down the column and contribute to the image. Its diameter therefore controls the ultimate

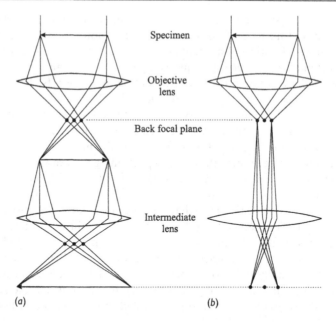

Specimen

Objective
lens

Back focal plane

Intermediate
lens

(a)   (b)

Figure 4.5 The objective and first intermediate lenses. The objective lens is focused on the specimen and forms an intermediate image as shown in (a). In imaging modes (a) the
intermediate lens magnifies this image further and passes it to the projector lenses
for display. In order to make the diffraction pattern visible (b), the intermediate
lens is refocused on the back focal plane of the objective lens and the diffraction pattern is passed to the projector system. An animated version of this diagram together
with a ray diagram can be found in MATTER: Introduction to electron microscopes.

resolution set by equation 1.7 and for high resolution a large aperture will be
needed. A modern 200 kV microscope, with a spherical aberration coefficient
C_s, of 1·2 mm, might show a resolution of 0·2 nm with an aperture of half-angle
4.5×10^{-3} radians (about a quarter of a degree).

However, for many purposes the highest resolution is not needed and the
objective aperture serves a different function in controlling the contrast which
will be seen in the image. In section 4.2 we consider the main contrast mechanisms available in a TEM.

4.1.5 The projector system – images

The first image produced by the objective lens usually has a magnification of
50–100 times. This is further magnified by a series of intermediate and *projector*
lenses and is finally projected onto the fluorescent screen. By using three or
four lenses, each providing a magnification of up to twenty times, a total
magnification of up to one million is easily achieved. It is not necessary to

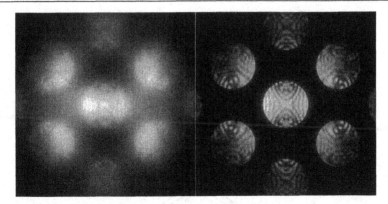

Figure 4.6 A convergent beam electron diffraction (CBED) pattern from relatively thick Si, taken with the incident beam along the [110] axis. Inelastic scattering gives a diffuse background and loss of contrast in the unfiltered image (left). An energy filter, tuned to accept only electrons which have been elastically scattered (zero energy loss electrons), removes this effect and gives a far clearer image (right). (Courtesy of C. B. Boothroyd, University of Cambridge)

use all the lenses to achieve a low magnification and in this case one or more projector lenses will be switched off.

Some specialized microscopes have an *energy filter* below the specimen, which can be tuned to allow the passage of only elastically scattered electrons or electrons which have suffered a particular energy loss. This has distinct advantages in, for example, high resolution electron microscopy (see section 4.2.4), since inelastic scattering degrades image quality. It is also useful in the quantitative interpretation of diffraction pattern intensities, since there is always a diffuse background of inelastically scattered electron intensity in an unfiltered diffraction pattern. Energy filters can also be deployed below the camera as part of an electron energy loss spectrometer. They can give dramatic improvements in the quality of images and diffraction patterns, especially from thick specimens where inelastic scattering effects are significant (Figure 4.6).

4.1.6 The projector system – diffraction patterns

It is often useful to examine or record the diffraction pattern from a selected area of the specimen. There are two fundamentally different ways of doing this. In the *selected area diffraction* technique, an area of specimen (usually circular) is selected, although a larger area is being illuminated. In the alternative *convergent beam diffraction* (also known as *microdiffraction*) technique the beam is condensed into a small spot so that the diffraction pattern comes from the whole of the (small) illuminated area.

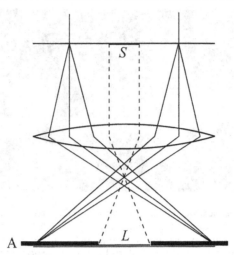

Figure 4.7 A selected area aperture A which selects a large area *L* in the intermediate image is optically equivalent to selecting the much smaller area *S* at the specimen.

Selected area diffraction could be achieved by inserting an aperture in the plane of either the specimen itself or the first image produced by the objective lens. The first of these is mechanically almost impossible while the second would require a very small aperture which is difficult to make, position and keep clean. The usual solution is to select the area using an aperture lower down the column, in the plane of the later intermediate images. In this way a much larger aperture can be used. For example if used in the plane of an intermediate image with magnification 200 ×, a 200 μm diameter aperture will select a region only 1 μm across. Figure 4.7 shows that selecting the area in an intermediate image is optically equivalent to selecting the smaller area at the specimen plane. The aperture used for this purpose is called the *selected area aperture*, or more loosely, the *diffraction aperture*. It is only useful to select an area 0·5–1 μm in diameter since one of the effects of spherical aberration is that electrons passing through the specimen as much as 1–4 μm outside the selected region may contribute to the diffraction pattern (depending upon the spherical aberration of the microscope). This is a small error if the selected region is 50 μm in diameter, but becomes the dominant feature if the selected area is only 1 μm in diameter.

The only way to obtain a good diffraction pattern from a region smaller than about 1 μm in diameter is to use the convergent beam diffraction technique (see section 3.4.2). In this case, the diameter selected is the same as the diameter of the beam at the specimen, which is controlled by the condenser lens system. In a modern microscope it may be possible to focus the electron beam (and hence selected areas) down to 1 or 2 nm in diameter, although in older instruments the limit may be as high as 100 nm.

4.1.7 The camera

Traditionally the camera has simply been a means of introducing a sheet of fine grain photographic film under the viewing screen, and a shutter mechanism (usually in the microscope column) to expose it. As described in section 1.7, the depth of focus is very large in a TEM due to the high magnifications used, and there is ample space for a variety of image capture systems which can be introduced at different positions under the viewing screen. With the increase in the availability and sophistication of digital imaging technology, charge-coupled device (CCD) array cameras are becoming more widespread. The simplest digital cameras simply use a computer-based video system directed at a phosphor screen. These have limitations due to the low light level from a phosphor screen and their video-standard resolution. Much more sophisticated cameras are also available which can capture images thousands of pixels wide. These cameras also measure the light from a phosphor screen, but in this case the light is channelled onto the CCD by a block of optical fibre. The CCD array is often cooled to reduce noise and longer exposure times are possible. Also, by combining the output from such a camera with a computer which has image processing software and control of the microscope lenses, it is now possible to perform some alignment procedures quickly and automatically.

4.1.8 Below the camera

Many analytical microscopes have electron detectors below the camera. These may be scintillation counters, which give an output proportional to the beam current and so can be used for STEM imaging (see section 4.4), or electron energy loss spectrometers, which give information about the effects the electrons have suffered on their passage through the specimen (see section 6.6), and in some cases allow energy-filtered images to be obtained (e.g. Figure 4.6).

4.1.9 Alignment

One of the most important aspects of good electron microscopy is the alignment of the electron beam along the optical axis of each lens. It is only if this alignment is accurate that the aberrations discussed in Chapter 1 can be minimized and the potential resolution of the microscope realized. Clearly the electromagnetic lenses must themselves have been mechanically well aligned, and this will have been done by the manufacturer. However, there is always the need for minor adjustments, which are performed by the operator using small electromagnetic deflection coils placed at strategic points in the column.

Alignment procedures differ from microscope to microscope but are always aimed at making sure that the beam is directed along the optical axis, particularly in the sensitive region near the objective lens. Further coils permit small fields to be imposed to correct for the effect of astigmatism in the condenser, objective and projector systems. All these adjustments need to be made by the operator before he or she attempts to take high magnification micrographs. The details of the procedures are beyond the scope of this book but can be found in textbooks such as that by Williams and Carter (1996).

4.2 Contrast mechanisms

All our discussion of light microscopy and TEM has so far been in terms of *geometrical optics*. Ray diagrams have been used to show that if there is an *object* (the specimen) in a certain plane, there will be a corresponding *image* at another plane. The viewing screen or camera is placed in the image plane and does not have to be positioned very precisely because the depth of focus is usually very large (section 1.7). Geometrical optics is very useful for considering the magnification, depth of field and focal conditions of a microscope but is virtually useless for interpreting the resultant image. For this purpose we must consider in detail the possible contrast mechanisms which will determine whether a feature will appear bright or dark in an image.

There are three basic contrast mechanisms, one, two or all three of which may contribute strongly to the appearance of a TEM image. In each case the important consideration is that the final image can only be formed using those electrons which pass through the objective aperture shown in Figure 4.5. Electrons stopped by the aperture will not contribute to the image. Clearly the size and position of the objective aperture are crucial in determining the nature of the contrast seen in the image, and this will be evident as we consider each mechanism in turn.

4.2.1 Mass–thickness contrast

If a specimen is thin enough to form a useful image in the TEM we can assume that, except for a few which are backscattered, primary electrons which enter the top of the specimen will emerge from the bottom. However, their angular range and energy spread will have been affected by elastic and inelastic scattering within the specimen. The effect of an aperture in the back focal plane of the objective lens is thus to stop all electrons which have been scattered, by any mechanism, through an angle greater than α in Figure 4.8. If the aperture is centred about the optical axis, as shown in Figure 4.8, then in the absence of a specimen a bright background is seen. This is known as *bright field* imaging. Regions of specimen which are thicker, or of higher density, will scatter more strongly (i.e. more electrons will be deflected through an angle greater than α) and will appear darker in the image. The effect is shown schematically in Figure 4.9. The mass–thickness contrast mechanism is exploited by most biological microscopists, who deliberately stain the thin specimen with a heavy metal, such as osmium, which

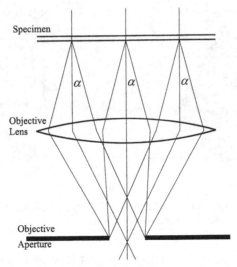

Figure 4.8 The objective aperture is located in the back focal plane of the objective lens. It prevents any electrons which have been scattered by the specimen through an angle larger than α from passing down the column and contributing to the image. An animated version of this diagram can be found in MATTER: Introduction to electron microscopes.

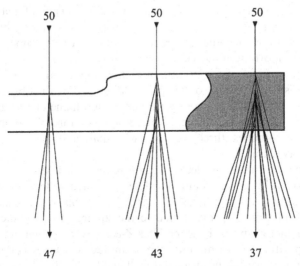

Figure 4.9 The scattering of electrons from different regions of a thin specimen. In a thin area (to the left) only a few electrons are scattered and perhaps 47 of the original 50 incident electrons continue undeflected. In a thicker region of the same material (centre) more are scattered and perhaps only 43 remain in the undeflected beam. From a region of the same thickness but higher density even more scattering will take place and perhaps only 37 electrons continue to pass through the objective aperture.

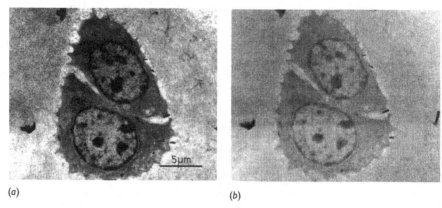

(a)   (b)

Figure 4.10 An animal cell (a cartilage cell from chick trachea) photographed (*a*) with and (*b*) without an objective aperture in position.

decorates specific features of interest. Figure 4.10 shows an animal cell which has been stained so that osmium has collected in the chromatin. If the objective aperture is removed (Figure 4.10(*b*)), the contrast almost disappears, since virtually all of the electrons can then contribute to the image. A little contrast remains because the bore of the microscope (only 1 mm or so in diameter) acts as a large aperture.

All transmission electron microscopists must be aware of mass–thickness contrast since all specimens – amorphous or crystalline, biological or inorganic – will show it. Other contrast mechanisms can be superimposed and may be more important in individual cases.

Both elastic and inelastic scattering of high-energy electrons only change the direction by a small amount, so the angle α subtended by the objective aperture often needs to be less than one degree. The actual diameter of the aperture is determined by the focal length (and hence magnification) of the objective lens; for modern microscopes, three apertures with diameters of 20, 50 and 100 µm are typically available.

In scanning transmission electron microscopes (STEMs), the same effect may be exploited by using an annular detector which collects electrons which have been scattered through relatively large angles. Since the inelastic scattering power of an atom is proportional to the square of its atomic number Z, atoms with higher Z appear brighter. This *Z-contrast imaging* is a useful way of getting compositional information from complex structures (Figure 4.11(*a*)). Furthermore, if a crystalline specimen is aligned such that the electron beam passes down a crystallographic axis and the probe size is sufficiently small, the inelastic scattering from each column of atoms can be imaged. These images have advantages over other high resolution images of atom columns in that the atom columns always appear bright, and appear brighter if they have a higher average atomic number (Figure 4.11(*b*)).

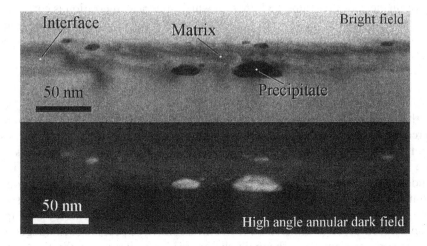

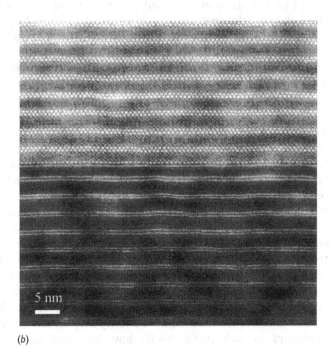

(b)

Figure 4.11 Atomic number (Z) contrast obtained using a high-angle annular dark field (HAADF) detector in a STEM. (a) Nickel precipitates in silicon appear dark in a bright field image and bright in a Z-contrast image. (b) Z-contrast image of a c-axis twist boundary in $Bi_2Sr_2Ca_2Cu_3O_{10}$. The position of the twist boundary between the double BiO layers can be identified directly from the Z-contrast image (courtesy of K. Kishida and N. D. Browning, University of Illinois at Chicago).

In the near future, 'super' STEMs, in which spherical aberration is corrected using complex computer-controlled lenses, will be able to achieve probe sizes smaller than 0·1 nm, which may allow investigation of materials on a sub-atomic scale.

4.2.2 Diffraction contrast – kinematical

If the specimen is crystalline then an additional contrast mechanism is encountered. Diffraction of the electron beam, described in detail in Chapter 3, means that the intensity of scattering is greatly increased at particular orientations of the specimen. The objective aperture can then be used to allow either the undeflected beam or a diffracted beam to form the image, thus giving strong contrast from regions which are diffracting strongly. There are many ways to use this contrast mechanism, and it is only possible to give a cursory description here. In order to understand the contrast which appears in such images, it is necessary to appreciate the main characteristics of electron diffraction. One of the more important parameters in the description of electron diffraction is the *extinction distance* (equation 3.9), and generally sharp images are only obtained when this is relatively small – a few tens of nm. Inspection of equations 3.10 and 3.11 shows that this only occurs for relatively low index reflections, and most diffraction contrast images are thus formed with the crystal oriented such that only the undeflected beam and one low index diffracted beam are present in the diffraction pattern.

Figure 4.12 illustrates the imaging conditions which are most frequently used. The sample is tilted so that a Kikuchi line runs through the undeflected beam (Figure 4.12(*a*)); the parallel Kikuchi line then runs through a strongly diffracted beam (section 3.4.1). In bright field imaging, the objective aperture is used to stop all diffracted beams and only permits undeflected electrons to contribute to the image (Figure 4.12(*b*)). If the aperture is displaced it can be used to select a particular diffracted beam, as shown in Figure 4.12(*c*). This is known as *dark field* imaging since in the absence of a specimen the background appears dark. However, if a dark field image is created by displacing the aperture, aberrations are likely to be introduced since all the imaging electrons are travelling far from the optical axis – where spherical aberration is large. A better method, avoiding this problem, is to tilt the incident electron beam so that the chosen diffracted beam travels along the optical axis and passes through a centred aperture (Figure 4.12(*d*)). This is technically simple to achieve but the microscopist needs to have a fairly deep understanding of the diffraction geometry to understand exactly how to orient both beam and specimen appropriately.

There is a wide variety of effects that can be achieved using diffraction contrast. Anything which changes the planes giving rise to Bragg diffraction becomes visible, including dislocations, stacking faults and other crystallographic defects. Changes in composition in multi-element compounds can

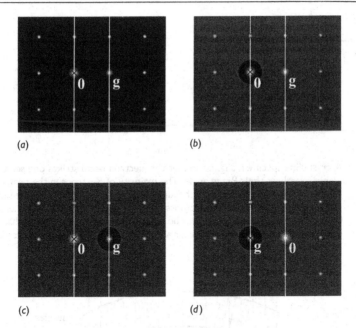

Figure 4.12 Common imaging conditions. (*a*) A diffraction pattern centred on the optical axis of the microscope (indicated by the cross). A two-beam condition (only one strong diffracted beam) has been produced by tilting the specimen such that a Kikuchi line runs through the undeflected beam. (*b*) The objective aperture centred over the undeflected beam to produce a bright field image. (*c*) The aperture displaced to permit a diffracted beam to pass, giving a low resolution dark field image. (*d*) Tilting of the incident beam so that the diffracted beam is on the optical axis, passing through the centred objective aperture, giving a standard dark field image. In reality the material around the aperture would of course be completely opaque. An animated version of this diagram together with a ray diagram can be found in MATTER: Introduction to electron microscopes.

alter the structure factor, which determines the diffracted intensity (section 3.1.2). The only disadvantage of this sensitivity occurs when the sample is bent. In this case the diffraction conditions will change across the field of view, which can make interpretation more difficult.

Before it is possible to understand the contrast which arises from defects, it is necessary to consider the effects which can arise with perfect crystals. A qualitative argument shows that a perfectly flat single crystal specimen can give rise to different intensity levels in the image depending on its exact orientation. Let us assume that one set of crystal planes is almost parallel to the electron beam and therefore close to fulfilling Bragg's law (equation 3.1). We will call these the *operating planes* and describe them by the reciprocal lattice vector **g**. A simplistic model (Figure 4.13) shows that if these planes are exactly at the Bragg condition strong diffraction will result and the

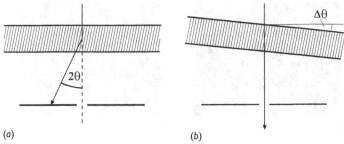

(a)   (b)

Figure 4.13 A crystalline specimen aligned so that the electron beam strikes one set of lattice planes exactly at their Bragg angle. The majority of electrons in the beam are diffracted and few continue to pass through the objective aperture. (b) The same specimen tilted a small angle $\Delta\theta$ from its previous position. Even if $\Delta\theta$ is as small as 0·25 degrees the lattice planes may be so far from the Bragg condition that a negligible amount of diffraction will occur and virtually the whole of the electron beam passes through the aperture.

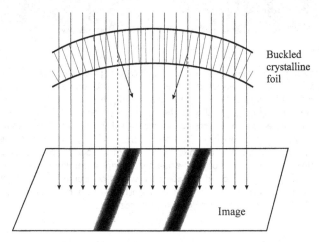

Figure 4.14 The bright field image of a buckled single crystal foil will show dark lines (extinction contours) in regions where the specimen is in the Bragg condition.

bright field image will therefore appear dark. If the specimen is tilted slightly (perhaps only a small fraction of a degree, exaggerated in Figure 4.13(b)), Bragg's law will no longer be obeyed, diffraction will be weak and a bright image will result.

Most real specimens are not perfectly flat and are elastically buckled to some extent. The image should therefore contain dark regions which correspond to those regions of the specimen which are at the Bragg angle, together with light regions corresponding to parts of the specimen which are not strongly diffracting. The dark regions are known as extinction contours and their formation is shown schematically in Figure 4.14. They are an inevitable feature of the

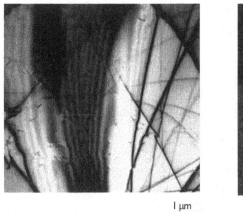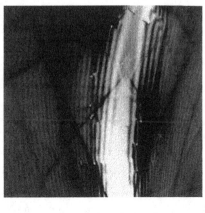

1 μm

(a)   (b)

Figure 4.15 A buckled region of a stainless steel foil which shows a major extinction contour (almost vertical) in both bright field (*a*) and dark field (*b*) imaging conditions. Notice the subsidiary fringes, which are especially clear in the dark field image.

contrast from a buckled crystalline specimen and their width serves to indicate the extent of buckling. Figure 4.15 shows a rather buckled region of a stainless steel foil containing many extinction contours, each of which arises from diffraction from a particular set of planes.

In Chapter 3 (equation 3.10) we discussed the expression for the intensity of diffraction as a function of thickness. Simplified slightly, this is:

$$I_g = \frac{\sin^2(\pi t s)}{(\xi_g s)^2} \qquad (4.1)$$

where s is the magnitude of the deviation parameter, which describes the distance in reciprocal space from the Bragg condition (section 3.3.2) and ξ_g is the extinction distance. This expression is only accurate for very thin specimens where the intensity of diffracted beams is negligible, in so-called kinematical conditions. However using this expression we can qualitatively interpret the fine details of extinction contours. A plot of I_g against s is known as a *rocking curve* since variation of s can be obtained by rocking a flat crystalline specimen through the Bragg condition. The rocking curve which is obtained from equation 4.1 at constant thickness t is shown schematically in Figure 4.16. Clearly the kinematical theory gives a strange result when s is close to zero since the intensity calculated from equation 4.1 then becomes very large. This is one of the reasons that the dynamical theory is needed and we therefore deal with it in the next section.

A buckled specimen provides a range of s values without the need for rocking, so we should expect extinction contours to show subsidiary fringes

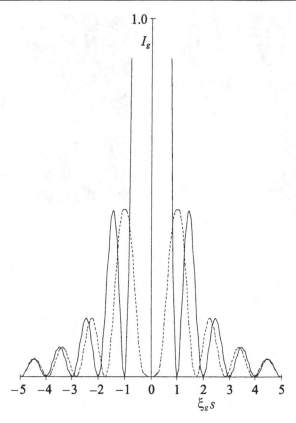

Figure 4.16 Rocking curves calculated for kinematical (full line) and dynamical (broken line) two-beam assumptions. The diffracted intensity is plotted against the parameter $\xi_g s$.

exactly as does the rocking curve. These fringes are visible on both sides of the main extinction contour in Figure 4.15. Extinction contours are one of the major imaging artefacts which it is necessary to be able to recognize in a crystalline specimen. They can be easily distinguished from actual crystal defects by tilting the specimen a little. Extinction contours, being simply contours showing the location of planes of similar orientation, will appear to move across the image as the specimen is tilted and different planes are brought into the Bragg condition. Defects embedded in the specimen will of course not move as the specimen is tilted, although they may change in appearance.

A second frequently observed image artefact can also be explained using equation 4.1. For a constant value of s (that is, for a fixed orientation) the intensity varies periodically with t, becoming zero each time the product ts is an integer. A typical wedge-shaped specimen, shown schematically in Figure 4.17(a), therefore shows thickness fringes which in a bright field image

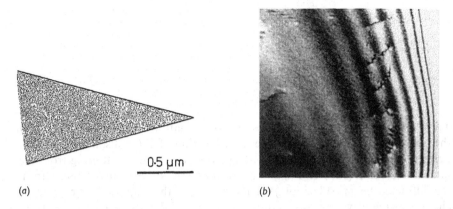

Figure 4.17 (a) Schematic diagram of the cross-section of a wedge-shaped specimen. (b) Thickness fringes (almost vertical) in a metal foil which gets thicker towards the left of the micrograph.

will be dark at the thickness $t = n/s$. Figure 4.17(b) shows the appearance of these fringes (see also Figure 4.42). Note that the kinematical theory cannot explain why the fringes still exist even at the exact Bragg orientation, when $s = 0$; the kinematical theory then predicts that intensity simply rises proportionally with thickness. We need the dynamical theory, outlined in the next section, to account for contrast at small values of s.

4.2.3 Diffraction contrast – dynamical

If a crystalline specimen is thicker than about one third of the extinction distance (ξ_g is a few tens of nm for most materials and operating reflections) then there will be appreciable interaction between the electron beams as they travel through the solid. For example, diffracted electrons can be scattered back into the original beam direction. In fact, this is very likely since a diffracted electron must be travelling at the Bragg angle to the diffracting planes and so is unlikely to get very far before it gets diffracted once more. This renders the kinematical assumption invalid and the dynamical theory is needed. The most straightforward version of this theory only considers the interaction between the undeflected beam and one diffracted beam defined by the reciprocal lattice vector **g**. This two-beam theory assumes that all other diffracted beams are very weak. Many-beam theories have been developed but they show that the main conclusions of the two-beam theory are not altered by the presence of other minor beams so we will not consider them here. There are several mathematical ways of formulating the two-beam theory, but the simplest to understand results in a pair of differential equations of this type:

$$\frac{d\phi_0}{dz} = \frac{i\pi}{\xi_0}\phi_0 + \frac{i\pi}{\xi_g}\phi_g \exp(2\pi isz)$$

$$\frac{d\phi_g}{dz} = \frac{1\pi}{\xi_g}\phi_0 \exp(-2\pi isz) + \frac{i\pi}{\xi_0}\phi_g \tag{4.2}$$

These Howie–Whelan equations describe the variation of the amplitudes, ϕ, of the undeflected and diffracted waves as a function of z, the distance through a perfect crystal. The first term in each equation arises from scattering from the undeflected beam (subscript 0) and the second term arises from scattering from the diffracted beam (subscript g). They show that the amplitude of each wave changes as the wave progresses through the crystal due to a contribution from the other.

This approach can be made even more realistic by including the possibility that electrons are 'absorbed' – or more accurately – inelastically scattered through a large angle and blocked by the objective aperture. This can be accounted for by replacing the (real) extinction parameter $1/\xi$ with a complex parameter $1/\xi + i/\xi'$. Although we shall not attempt to justify this approach in this short text, it is worth noting that equations 4.2 and their refinements are used to describe all diffraction contrast effects with very good experimental agreement for a wide variety of situations.

Equations 4.2 can be solved analytically for a perfect crystal. In order to calculate the intensity at any point in the image the equations must be integrated over the whole thickness to give ϕ_0 and ϕ_g at the exit surface of the specimen. The bright field intensity is then given by $\phi_0\phi_0^*$ and the dark field intensity by $\phi_g\phi_g^*$, where * indicates the complex conjugate. The resultant diffracted beam (dark field) intensity, for a perfect crystal of thickness t, is given by

$$I_g = \frac{\sin^2(\pi ts')}{(\xi_g s')^2} \tag{4.3}$$

which is identical to the kinematical solution (equation 4.1) except for the parameter s', known as the effective deviation parameter. The relationship between s' and s is

$$s' = \left[s^2 + \left(\frac{1}{\xi_g}\right)^2\right]^{1/2} \tag{4.4}$$

Equation 4.3 behaves reasonably even when s approaches zero and strong diffraction occurs. An accurate rocking curve can be deduced, as is shown in Figure 4.16. The presence of thickness fringes can also be understood more clearly – in a bright field image, a bright fringe occurs when most of the diffracted electrons have been diffracted back into the undeflected beam.

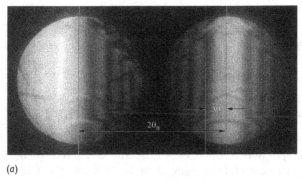

(a)

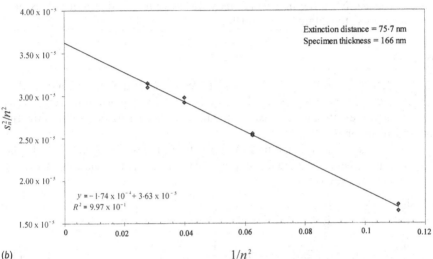

(b)

Figure 4.18 (a) A two-beam $\mathbf{g} = 220$ CBED pattern from a silicon single crystal. (b) The extinction distance and specimen thickness can be obtained from the slope of a plot of s_n^2/n^2 against $1/n^2$, where s_n is the value of s at the nth fringe.

The extinction distance is the thickness of the specimen at the first bright fringe.

Equations 4.3 and 4.4 can be used to measure both the extinction distance and the crystal thickness from a two-beam CBED pattern (Figure 4.18(a)). A 'rocking curve' is present in the dark field disk of such a pattern, since the relationship between the beam direction and crystal planes for a convergent beam and a flat specimen is similar to that of a parallel beam and a bent specimen.

From equation 4.3, a bright fringe occurs every time $\sin^2(\pi t s') = 0$, i.e.

$$\pi t \left[s_n^2 + \frac{1}{\xi_g^2} \right]^{1/2} = n\pi \qquad (4.5)$$

which can be rearranged to give

$$\frac{s_n^2}{n^2} = -\frac{1}{n^2 \xi_g^2} + \frac{1}{t^2} \tag{4.6}$$

Now, s_n (the s value at the nth fringe) is simply

$$s_n = \frac{\lambda}{d^2} \frac{\Delta\theta}{2\theta_B} \tag{4.7}$$

where $\Delta\theta$ and $2\theta_B$ are shown on Figure 4.18(a). From equation 4.6, a plot of s_n^2/n^2 against $1/n^2$ has a slope of $-1/\xi_g^2$ and a y-intercept of $1/t^2$. Figure 4.18(b) shows the calculation using data taken from the CBED pattern of Figure 4.18(a). Note the first fringe may not have $n = 1$; different values of n may be needed to obtain a straight line.

Defects

Crystal defects are often of interest and TEM is one of the few techniques that can characterize them in detail. The contrast in TEM images of defects is thus of considerable interest, and can only be explained satisfactorily using the dynamical theory.

A crystal defect which disturbs the operating planes will locally modify the deviation parameter. A more useful pair of Howie–Whelan equations (ignoring absorption for brevity) is

$$\frac{d\phi_0}{dz} = \frac{i\pi}{\xi_0} \phi_0 + \frac{i\pi}{\xi_g} \phi_g \exp[2\pi i(sz + \mathbf{g} \cdot \mathbf{R})]$$

$$\frac{d\phi_g}{dz} = \frac{i\pi}{\xi_g} \phi_0 \exp[-2\pi i(sz + \mathbf{g} \cdot \mathbf{R})] + \frac{1\pi}{\xi_0} \phi_g \tag{4.8}$$

In these equations $\mathbf{g}$ is the operating diffraction vector (i.e. the reciprocal lattice vector which describes the planes which are diffracting) and $\mathbf{R}$ is the displacement of atoms from their lattice positions due to the defect, which may vary with z.

In order to solve equations 4.8, a numerical computer method is generally necessary. It is usually assumed that the intensity at one point (x, y) in the image, which arises from one column $(x, y, z = 0$ to $t)$ in the specimen, is independent of the behaviour of the electron beams in adjacent columns. This is known as the *column approximation* and it holds very accurately in most situations. The computed image is built up pixel by pixel, as Figure 4.19 illustrates.

The term $\mathbf{g} \cdot \mathbf{R}$ in equations 4.8 simply modifies the product sz and can thus be interpreted in terms of the effect which the defect has on the operating planes. When $\mathbf{g} \cdot \mathbf{R}$ is zero, the displacements do not disturb the operating planes and so the defect is invisible. For many of the crystal defects of most

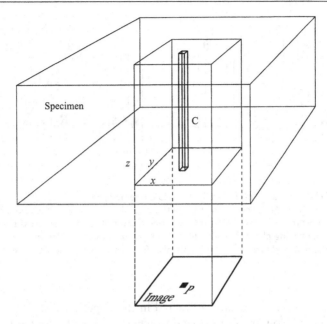

Figure 4.19 The geometry of a TEM image, either calculated or real. The image consists of pixels such as P, the intensity of which arises from scattering in the column C at position *x, y*.

interest we know the nature of the displacement field $\mathbf{R}(x, y, z)$ and so we can understand and predict the appearance of their images in the TEM. We will consider here a few simple examples.

Planar defects

Stacking faults, grain boundaries and phase boundaries are planar defects which are frequently studied in projection as they cross the thin specimen. For a stacking fault, if we ignore any local strain, then in each column $\mathbf{R}$ is effectively zero for all positions above the fault plane and has a non-zero but constant value everywhere below the fault. If the fault is a displacement fault, such as the common stacking fault in the fcc structure, then the value of s is the same above and below the fault, because no lattice planes are tilted by the fault. If the fault has a misorientation associated with it, then the value of s will change abruptly at the fault plane. Figure 4.20 shows these two types of behaviour. In some circumstances faults giving rise to both displacement and misorientation can occur; in these cases both $\mathbf{R}$ and s change at the fault.

The characteristic contrast of stacking faults consists of fringes running parallel to the intersection of the fault with the specimen surface. Figure 4.21 shows an example. The fringes can be thought of as resulting from the

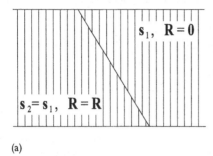 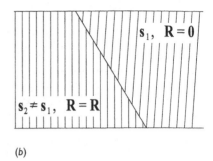

(a)   (b)

Figure 4.20 Schematic diagram showing the two extreme types of planar defect. (*a*) In a displacement fault the orientation of the crystal is identical on both sides of the fault and thus **s** is unchanged. (*b*) In an orientation fault **s** changes across the fault since the imaging planes are slightly tilted. In both cases the lattice may be displaced at the fault and thus in general **R** changes from zero to some finite value.

interference between the beams diffracted in the upper part of the crystal and those diffracted below the fault. The diffracted waves suffer a phase change at the fault and the fringes are thus the usual manifestation of interference between two sets of waves which are slightly out of phase.

Two aspects of fault contrast are of particular interest. It is clear from equations 4.5 that if the product $\mathbf{g} \cdot \mathbf{R}$ is zero (or indeed an integer) then the crystal containing a stacking fault will give contrast identical to that from a perfect crystal. The condition for $\mathbf{g} \cdot \mathbf{R}$ to be zero is that the fault vector **R** is perpendicular to the operating reflection **g**. Thus for example in Figure 4.21(*c*) the faults are invisible because the fault vector is 1/6[112] and **g** is $2\bar{2}0$. Another way of thinking about this is that the image is being formed using diffraction from the $(1\bar{1}0)$ plane but this plane is unaffected by the presence of the stacking fault. The fault thus cannot be detected using these diffraction conditions. This '*invisibility criterion*' can be used to identify the displacement vector of a stacking fault.

A second useful form of contrast can be exploited by imaging a displacement fault in both bright field and dark field. Although the kinematical theory would imply that the bright field and dark field images should be complementary (i.e. with dark and bright fringes reversed), this is not what is found in reasonably thick specimens. The dynamical theory correctly predicts the contrast illustrated in Figure 4.21(*a*) and (*b*), which shows a top/bottom effect. The calculated fringe image is shown in Figure 4.22. The images from the region of fault near the bottom of the specimen are indeed complementary but the images from the top of the foil are similar (i.e. a bright fringe in dark field corresponds to a bright fringe in bright field). Thus the two possible inclinations of the fault can be distinguished, as Figure 4.21 shows.

(a)

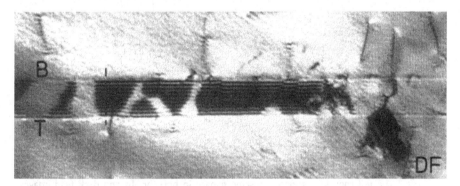

(b)

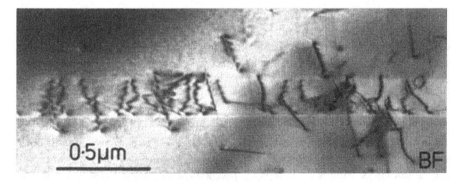

(c)

Figure 4.21 Three images of a heavily faulted region in a specimen of stainless steel. In (a) and (b) the faults are in contrast while in image (c) they are invisible. Images (a) and (b) are a bright field/dark field pair, from which the orientation of the fault may be deduced. The calculated image from a similar fault is shown in Figure 4.19. The complementary nature of the fringes at the bottom (B) of the specimen and the similarity at the top (T) is particularly clear at A.

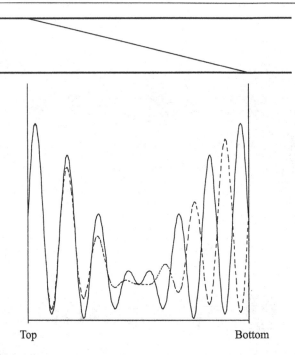

Top   Bottom

Figure 4.22 Calculated fringes from a stacking fault with $\mathbf{g}\cdot\mathbf{R} = 1/3$ in a specimen of thickness $7\xi_g$. The bright field image (full line) is symmetric about the middle of the specimen while the dark field image (broken line) is not. The part of the fault at the top of the foil therefore shows the same fringes in both bright and dark field images, while at the bottom of the foil the bright field and dark field images are complementary.

 The other main type of planar defect is the grain or phase boundary, across which the orientation (and possibly the lattice parameter or structure) change significantly. In most cases this means that we cannot describe the diffraction conditions on both sides of the boundary in terms of the same $\mathbf{g}$. One grain, say the top grain in Figure 4.23, may be set up with two-beam conditions characterized by a deviation s_1 from the operating reflection $\mathbf{g}_1$. The lower grain may be unlikely to be near a two-beam condition, and even if it is the values of $\mathbf{g}_2$ and s_2 will be different from $\mathbf{g}_1$ and s_1. The most common consequence of this is that the strongly diffracting grain (top in this example) shows thickness fringes as if it were just a tapered single crystal, while the other (bottom) grain contributes little to the contrast. The net effect is a set of fringes running parallel to the intersection of the boundary with the surface (Figure 4.24). The spacing of the fringes depends on s_1. A dark field image will allow these fringes to be easily distinguished from those of a stacking fault, since the diffracting grain will appear bright and the grain which is not diffracting will appear dark.

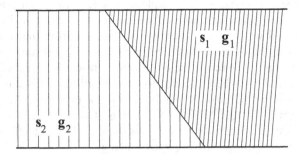

Figure 4.23 The diffraction conditions at a grain boundary. If the upper (right) grain is oriented such that the operating planes deviate by s_1 from $\mathbf{g}_1$, then it is likely that a different vector ($\mathbf{g}_2$) will be excited in the lower (left) grain, at a different and probably much larger deviation s_2.

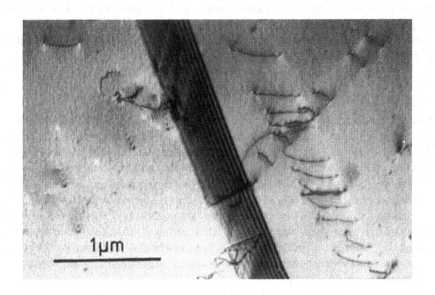

Figure 4.24 Grain boundary fringes in a stainless steel specimen.

When both crystals are strongly diffracting, yet another source of fringes can appear. These are *moiré* fringes. They are common in images of thin layers of one crystalline material deposited on another, but can arise in any situation where two crystals are diffracting with slightly different values of $\mathbf{g}$. The situation is illustrated in Figure 4.25(*a*), where the planar boundary is parallel to the plane of the specimen, and the lower crystal has a slightly larger lattice parameter than the upper crystal. We ignore any differences in

ξ_g and s in the two crystals. The displacement $\mathbf{R}$ varies with position in the crystal below the boundary plane. If we choose one point to have zero displacement, then $\mathbf{R}$ will increase linearly with distance from that point, and will point away from the origin. Since it is $\mathbf{g} \cdot \mathbf{R}$ which is important for changes in the electron intensity, there will be a line, along which $\mathbf{R}$ is perpendicular to $\mathbf{g}$ and hence $\mathbf{g} \cdot \mathbf{R}$ is zero, where the electron beam will have the same intensity as at the origin (Figure 4.25(b)). Perpendicular to this direction, the intensity will also be the same as at the origin every time $\mathbf{g} \cdot \mathbf{R}$ is an integer. Between these points the electron amplitude varies sinusoidally. The net effect is a set of fringes running perpendicular to $\mathbf{g}$ with spacing

$$d = \frac{1}{g_1} - \frac{1}{g_2} \tag{4.9}$$

Moiré fringes can also be seen at boundaries between crystals with a slight rotation. The only condition is that a strongly diffracted beam from each crystal passes through the objective aperture. An important point is that the fringes will be seen in both bright field and dark field images. An example of moiré fringes seen at the interface between Pd_2Si and silicon is shown in Figure 4.25(c).

Line defects – dislocations

Near the core of a dislocation, lattice planes are usually bent quite severely but the extent of lattice bending decreases at greater distances. Qualitatively it is quite easy to explain the appearance of an edge dislocation as a dark line in the TEM (e.g. Figure 4.26). Figure 4.27 shows lattice bending schematically. If the crystal far from the dislocation is set in a two-beam condition, near to but not at the Bragg orientation (i.e. s is not zero), then the bent planes on one side of the dislocation core must reach the Bragg orientation ($s = 0$) and will diffract more strongly than their surroundings. This is true all along the dislocation line and it will therefore appear as a dark line in a bright field image. A similar argument can be made for a screw dislocation, although the plane bending is harder to show in a sketch.

In terms of equation 4.5 the displacement $\mathbf{R}$ is clearly a function of x, y and z for a dislocation. $\mathbf{R}$ therefore varies along each column in an image (Figure 4.28) and it is necessary to integrate the Howie–Whelan equations numerically in order to calculate the image contrast from a dislocation. It is evident however that if $\mathbf{g} \cdot \mathbf{R}$ is zero all the way down a column the dislocation must be invisible. The displacement field of a dislocation in an infinite isotropic solid is well known and can be written as

$$\mathbf{R} = \frac{1}{2\pi} \left\{ \mathbf{b}\phi + \mathbf{b}_e \frac{\sin 2\phi}{4(1-v)} + \mathbf{b} \times \mathbf{u} \left(\frac{1-2v}{2(1-v)} \ln r + \frac{\cos 2\phi}{4(1-v)} \right) \right\} \tag{4.10}$$

where r and ϕ are defined in Figure 4.28 and v is Poisson's ratio for the

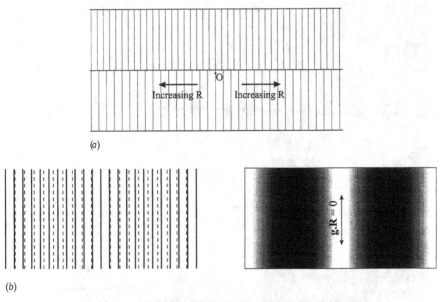

(a)

Increasing R   Increasing R

O

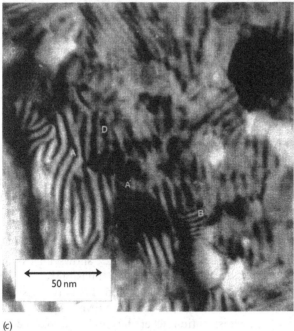

(b)

g.R = 0

(c)

Figure 4.25 (a) The case of a boundary between dissimilar materials in the plane of the specimen. The displacement vector R increases away from the origin O. (b) The formation of moiré fringes when the operating planes in the two crystals are parallel. (c) An image showing moiré fringes in a specimen of Pd_2Si on silicon. (N. A. McAuley)

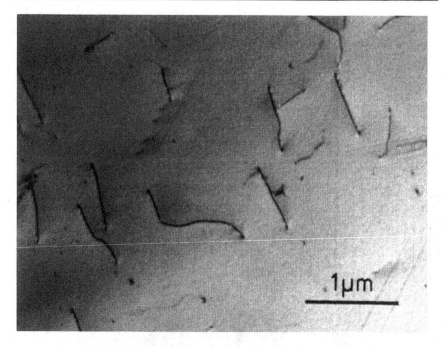

Figure 4.26 Dislocations in strong diffraction contrast in a metal specimen.

material; **b** is the Burgers vector and $\mathbf{b}_e$ is its edge component (i.e. that component perpendicular to the dislocation line vector **u**). For a screw dislocation, equation 4.7 simplifies to

$$\mathbf{R} = \frac{\mathbf{b}\phi}{2\pi} \tag{4.11}$$

In this case, **R** is directly proportional to **b** and therefore such a dislocation will be invisible if $\mathbf{g} \cdot \mathbf{b}$ is zero. For example, a screw dislocation of Burgers vector $a/2[110]$ will be invisible in a two-beam image with $\mathbf{g} = 002$ or $\bar{1}11$ but visible when $\mathbf{g} = 202$ (when $\mathbf{g} \cdot \mathbf{b} = 1$). Another way of putting this is that this dislocation does not bend the (002) plane but it does bend the (101) plane (which is the same as bending the (202) plane). Figure 4.29 shows an example of screw dislocation invisibility. If two conditions of invisibility can be found the direction of **b** can be determined uniquely and this is the basic technique for finding the Burgers vectors of dislocations in TEM images.

For an edge or mixed dislocation, where **b** is no longer parallel to **u**, equation 4.6 shows that $\mathbf{g} \cdot \mathbf{R}$ is only zero when $\mathbf{g} \cdot \mathbf{b}$ and $\mathbf{g} \cdot \mathbf{b} \times \mathbf{u}$ are both zero. This is a more difficult condition to satisfy and very often an edge dislocation will display some faint residual contrast when $\mathbf{g} \cdot \mathbf{b} = 0$ because $\mathbf{g} \cdot \mathbf{b} \times \mathbf{u}$ is not also zero.

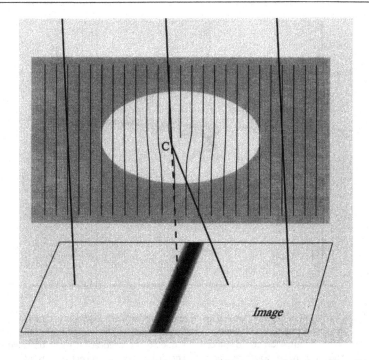

Figure 4.27 Diffraction contrast at an edge dislocation. If the crystal far from the dislocation (shaded area) is set at an orientation close to the Bragg condition (i.e. with **s** large), then the lattice on one side of the dislocation core at C will be bent locally into the Bragg condition (**s** = 0). This part of the specimen will therefore diffract the beam strongly and will appear dark in a bright field image.

Since equation (4.7) applies to dislocations in infinite material and the TEM specimen is very thin, we might expect some differences in the contrast behaviour of dislocations in regions where relaxation of the internal strain is significant – i.e. at the surfaces of the specimen. This is visible in Figure 4.29, where dislocations invisible in the bulk of the specimen show a dot of contrast where they meet the specimen surface.

It is generally assumed that for lattice dislocations, where $\mathbf{g} \cdot \mathbf{b}$ must be an integer, dislocations are invisible if $\mathbf{g} \cdot \mathbf{b} = 0$, visible if $\mathbf{g} \cdot \mathbf{b} = 1$ and visible but likely to have unusual images (e.g. a double line) if $\mathbf{g} \cdot \mathbf{b} \geqslant 2$. Partial dislocations may however have non-integral values of $\mathbf{g} \cdot \mathbf{b}$. For example, a Shockley partial dislocation with $\mathbf{b} = a/6[112]$ would have $\mathbf{g} \cdot \mathbf{b} = 1/3$ if imaged with $\mathbf{g} = 200$ and $\mathbf{g} \cdot \mathbf{b} = 2/3$ using $\mathbf{g} = 220$. The rule of thumb for visibility is that a partial dislocation is visible if $\mathbf{g} \cdot \mathbf{b} > \frac{1}{2}$. However it should be borne in mind that, by definition, a partial dislocation must bound a region of stacking fault, so it will be necessary to consider the visibility of both the stacking fault and the dislocation in order to interpret an image containing partial dislocations. In

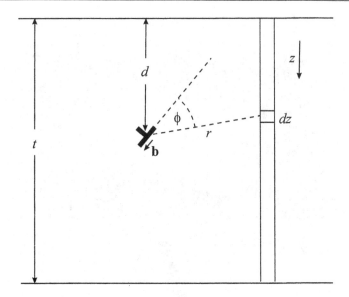

Figure 4.28 A column of crystal close to an edge dislocation with Burgers vector **b** at a depth *d* in a specimen of thickness *t*. For each step in the integration of the Howie–Whelan equation the displacement due to the dislocation must be calculated for the element *dz* positioned at *r*, φ. The dislocation line direction **u** is into the paper.

Figure 4.21 for example, most of the partial dislocations bounding the faults are out of contrast in images (*a*) and (*b*) but are in very strong contrast in image (*c*) when the faults themselves are invisible.

It is increasingly the case that we know the nature of most defects in materials of common crystal structure. The problems which are now being addressed tend to involve dislocations in complex crystal structures, new alloys and materials, and in unusual combinations. For these purposes the simple 'two invisibility conditions' method of determining **b** is often unsatisfactory and an alternative approach is needed. One approach is to compute the images expected from likely dislocation geometries and to compare these calculated images with experimental images taken using the appropriate diffraction conditions. An example of this technique is shown in Figure 4.30. Each computed image consists of about 100 000 pixels; the intensity has been calculated by numerically integrating the Howie–Whelan equations using the column approximation. Another approach is to use the dynamical diffraction effects which occur when a relatively high index bend contour crosses a dislocation image (Figure 4.31). It has been shown that a bend contour, which is a single line away from the dislocation, splits into braids when it crosses the dislocation line. The number of braids gives the value of **g·b**, and the shape of the braiding can be used to determine whether **g·b** is positive or negative.

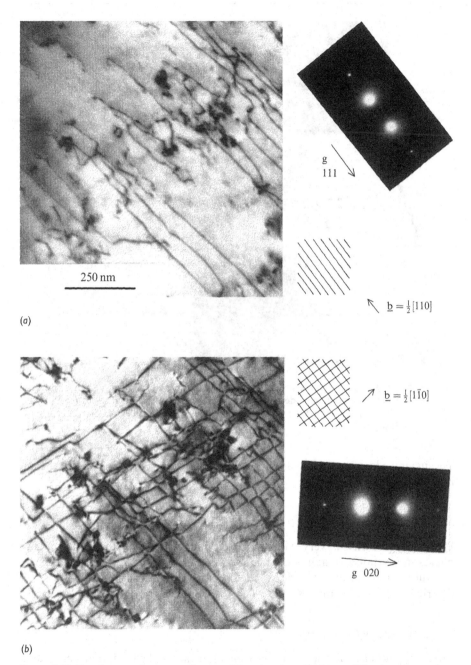

250 nm

(a)

g
111

$\underline{b} = \frac{1}{2}[110]$

$\underline{b} = \frac{1}{2}[1\bar{1}0]$

g 020

(b)

Figure 4.29 Two images of an array of screw dislocations in a niobium-stabilized steel, together with the diffraction pattern in each case. In (a) one set of dislocations is out of contrast since its Burgers vector is perpendicular to the operating **g** vector. In (b) both sets are visible.

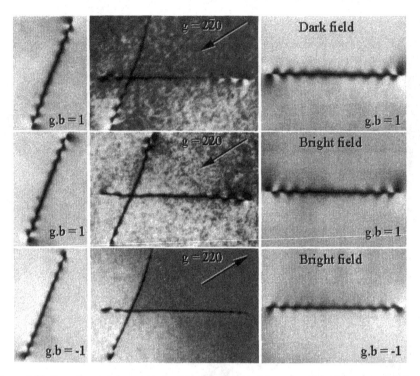

Figure 4.30 Experimental (centre column) and computed (outer columns) images of disloca-
tions in gallium arsenide under three different diffraction conditions (courtesy of
Drs Ruvimov and Scheerschmidt, Max Planck Institut für Microstrukturphysik,
Halle).

The magnitude of $\mathbf{g} \cdot \mathbf{b}$ can also be determined from dark field, weak beam
images (see below).

Some other examples of diffraction contrast

Diffraction contrast is very widely used to image specific defects and second
phases in crystalline materials. Once the essential relationship between image
and diffraction pattern is understood, the possibilities are limitless. We shall
describe just a few examples.

SMALL CRYSTALLINE PARTICLES

These may differ from the matrix in which they are embedded in atomic mass,
lattice parameter, crystal structure or orientation. They may also strain the
matrix in rather the same way as a dislocation, bending lattice planes. Any
one of these effects could be used to form an image of the particles and their

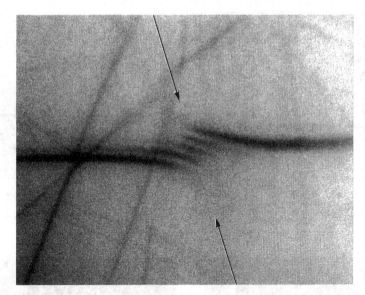

Figure 4.31 Braiding of a 440 extinction contour in a large-angle CBED pattern as it crosses a 1/2[110] dislocation (arrowed) in a diamond specimen. The number of fringes gives $g \cdot b + 1$. (Courtesy of D. Cherns, University of Bristol.)

distribution. For example one of the small precipitates in Figure 4.32(*a*) can be seen because it is of different structure to the matrix and diffracts strongly in these imaging conditions. In Figure 4.32(*b*) the same strained particle, together with another, can be seen because the matrix is elastically strained and the imaging lattice planes are bent. The images are therefore rather larger than the particles themselves. The larger particles in Figure 4.32(*c*) can be seen by virtue of their higher density, while in Figure 4.32(*d*) some of them have been made to show up in a dark field image formed by permitting only the electron beam resulting from diffraction within appropriately oriented particles to pass through the objective aperture.

A SMALL CAVITY

This is an extreme case of a local change in density of the specimen. The image of a cavity may therefore arise from mass–thickness contrast if kinematical conditions prevail (i.e. small thickness or large value of **s**). The cavity will then appear lighter than its background in a bright field image, although this effect may be small and a more noticeable effect may be the Fresnel fringes arising from near field diffraction at the interfaces when the image is defocused. However in dynamical conditions the column of crystal containing the cavity is of appreciably smaller thickness than its surroundings and the cavity may show

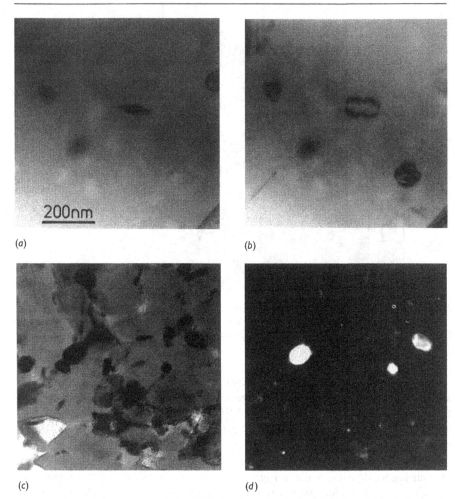

Figure 4.32 Precipitates in a rapidly solidified and extruded Al-Cr-Zr alloy. (*a*) Matrix weakly
diffracting and one particle strongly diffracting; (*b*) the same area as (*a*) with the
matrix set close to a strong diffracting condition. Two particles show strain contrast. (*c*) A bright field image showing large particles dark because of absorption.
(*d*) the same area as (*c*) imaged in dark field using a precipitate spot to show some
of the particles clearly. (P. S. Goodwin and N. J. E. Adkins, University of Surrey)

contrast arising from thickness fringes. Examples of these various types of
contrast are shown in Figure 4.33.

DARK FIELD, WEAK BEAM IMAGING CONDITIONS

These can be used to observe fine detail in images of dislocations, and under
appropriate conditions, determine the magnitude of **g·b**. A rough estimate of

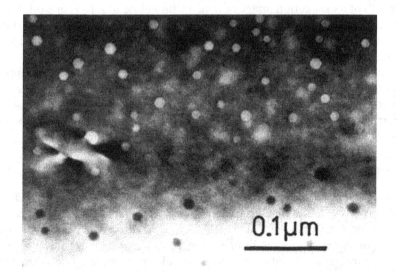

(a)

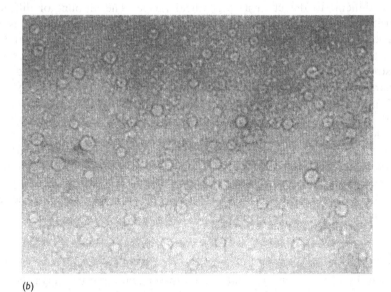

(b)

Figure 4.33 Two images of the same region of an aluminium specimen containing small helium bubbles. (a) In dynamical conditions the cavities appear bright against dark regions of the extinction contour and dark against bright regions. (b) In kinematic conditions the cavities are very slightly brighter than the background but are more visible when, as here, the image is very slightly out of focus and each bubble is surrounded by a dark fringe.

the width of a dislocation image obtained with s close to zero is one third of the extinction distance ξ_g for the operating reflection $\mathbf{g}$. Since extinction distances for low index reflections for most crystalline materials are in the range 10–200 nm, the image width is appreciable, and may make it difficult to resolve dislocations which are close together. This situation can be greatly improved using the 'weak beam' technique. The dynamical theory of electron diffraction showed that the effective deviation parameter $\mathbf{s}'$ is given by equation 4.4. In a similar way it can be shown that the effective extinction distance when $\mathbf{s}$ is not zero is:

$$\xi_{\text{eff}} = \frac{1}{s'} = 1/\{s^2 + (1/\xi_g^2)\}^{-1/2} \tag{4.12}$$

When s is large, dislocation images have a width of approximately $\xi_{\text{eff}}/3$. At a deviation of $s = 5/\xi_g$ (which is easily attainable) the image width will be about a factor of 5 smaller than at $s = 0$. Figures 4.34(a) to (c) show images of dislocations in Si for the conditions s close to zero and $s = 25/\xi_g$. The improvement is obvious. Note also that in the weak beam image, thickness fringes terminate at the ends of the dislocation to the right. The number of fringes which terminate gives the value of $\mathbf{g} \cdot \mathbf{b}$ for the reflection used.

Weak beam images are usually taken in dark field since the contrast is small, and is difficult to detect in a bright field image. The amount of diffracted intensity in the 'weak' beam with diffraction vector $\mathbf{g}$ is improved by using a stronger diffracted beam, usually with diffraction vector $2\mathbf{g}$ or $3\mathbf{g}$, and the diffraction conditions are usually described as '$\mathbf{g}$, $N\mathbf{g}$', where N describes the strongly diffracted beam. The larger the value of N, the larger is s and the narrower the image. Obviously, a limit is reached in that the intensity of the weak beam becomes so low it becomes difficult to obtain any image; even images taken using $\mathbf{g}$, $3\mathbf{g}$ imaging conditions may need exposures five to ten times longer than the bright field image.

The way a $\mathbf{g}$, $3\mathbf{g}$ imaging condition is set up using the diffraction pattern is shown in Figure 4.34(d). Again, the diffracted beam of interest is made to lie on the optical axis of the microscope by tilting the incident beam. As is apparent from comparison with Figure 4.12, in the case of weak beam imaging the sense of tilt is opposite to that used for conventional dark field imaging.

COMPOSITIONAL CONTRAST

This can be obtained in materials which the crystal structure stays the same while the composition varies (i.e. solid solutions). If a diffracted beam, which has a structure factor dependent upon the composition of the material, is used to form an image changes in composition can easily be detected. For example, in ternary or quaternary III–V semiconductors, the 002 diffracted beam has an intensity which is roughly proportional to the difference in atomic number, Z, between the group III and group V elements (ignoring absorption effects). An

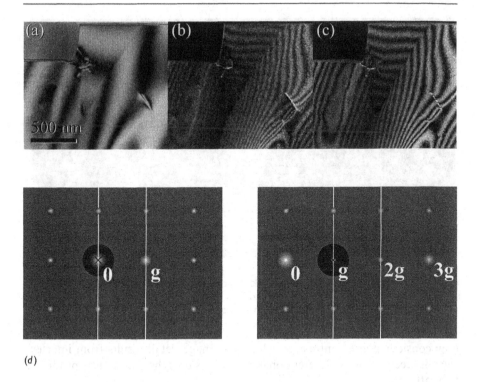

Figure 4.34 (*a*) A bright field and (*b*), (*c*) dark field, weak beam images of dislocations in silicon. Note the decrease in the width of the dislocation image and the termination of thickness fringes at the ends of the dislocation, ($\mathbf{g} \cdot \mathbf{b} = 2$ in (*b*), $\mathbf{g} \cdot \mathbf{b} = 1$ in (*c*)). (*d*) The diffraction pattern for a $\mathbf{g}$, $3\mathbf{g}$ dark field weak-beam imaging condition. Note that the incident beam is tilted in the opposite direction to that used for conventional dark field imaging (Figure 4.12).

example of such an image is shown in Figure 4.35, where the InP layers appear bright (average difference between group III and V atomic number $= 17$) and the $In_{0.53}Ga_{0.47}As$ layers appear dark (average difference between group III and V atomic number $= 4.77$).

Two important points should be emphasized concerning the contrast mechanisms discussed so far. Mass–thickness and diffraction contrast may be present at the same time, as for instance in Figure 4.24, in which the region at the lower left is thinner and therefore brighter than the rest. Secondly, several manifestations of the effect are likely to appear in any specimen. For example the micrographs in Figures 4.15, 4.18, 4.21 and 4.24 were all taken from neighbouring regions of the same specimen. The diffraction conditions were of course carefully selected by tilting the specimen or beam or both.

Both mass–thickness and diffraction contrast are amplitude contrast mechanisms because they employ only the amplitudes of the scattered waves. We

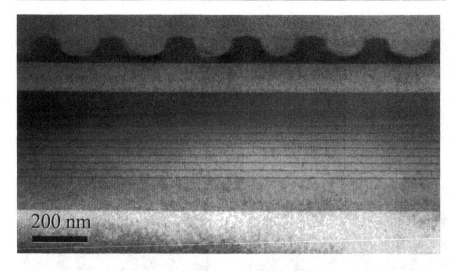

Figure 4.35 A dark field **g** = 002 image of layers of different compositions in a III–V hetero-junction laser structure. The contrast arises from the change in the structure factor of the 002 diffracted beam with composition.

now consider phase contrast, in which the image detail results from interference between waves of different phases and thus uses the phase shifts produced by scattering.

4.2.4 Phase contrast

Phase contrast results whenever electrons of different phase are allowed to pass through the objective aperture. Since most electron scattering mechanisms involve a phase change this means that some sort of phase contrast is present in every image, since it is impossible to use an objective aperture small enough to stop any deflected electrons contributing to the image. There therefore tends to be a phase contrast component to all images, sometimes noticeable at high magnifications as a speckled background. However phase contrast becomes really useful when two or more diffracted beams are allowed to pass through the objective aperture and interfere. In order to achieve this a larger aperture is needed than is commonly used for diffraction or mass–thickness contrast.

Each pair of beams which interferes will in principle give rise to a set of fringes in the image. The simplest way of exploiting this is to obtain a two-beam condition and allow just those two beams to pass through the aperture, as shown in Figure 4.36. A better alternative, avoiding spherical aberration effects, is to place the undiffracted beam halfway between the Kikuchi lines and use three beams: $+\mathbf{g}$, $\mathbf{0}$ and $-\mathbf{g}$. In both these cases a lattice image is formed, in

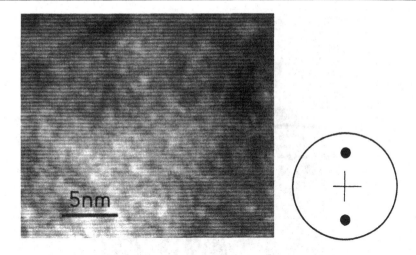

Figure 4.36 A two-beam lattice image of gallium arsenide showing one set of fringes from the (111) planes (horizontal). The two beams (000 and 111) were symmetrically placed around the optical axis, as shown in the diagram, in order to minimize spherical aberration. (U. Bangert, UMIST)

which each dark fringe appears to represent a lattice plane. This may be true in ideal conditions but for reasons beyond the scope of this book there is not in general a one-to-one correspondence between fringes and planes. Nevertheless many useful images have been recorded this way and some of the earliest micrographs of this type of image confirmed the idea of an 'extra half plane' at an edge dislocation.

A more common, and indeed more useful, type of phase contrast image is formed when more diffracted beams are used to form the image. By tilting the crystal such that the incident beam is accurately parallel to a zone axis in the crystal, many strong diffracted beams can be produced (e.g. Figure 3.14). Selecting several beams allows a *structure image* (often called a high-resolution electron microscope, or HREM, image) to be formed (Figure 4.37). The many lattice fringes intersect and give a pattern of dark (or bright) spots corresponding to atom columns. The interpretation of the contrast in these images is not straightforward, since the contrast in the image changes dramatically with specimen thickness and defocus, and is also strongly dependent upon the resolution of the microscope, which is itself dependent upon variables such as the spherical aberration coefficient C_s of the objective lens, and the wavelength of the electrons. It is possible that the same specimen in the same microscope can give an image in which one dark spot corresponds to columns of atoms, the spaces between them, or even a pair of atom columns.

Good HREM images can only be obtained from thin specimens, since inelastic scattering will tend to degrade the image quality. The microscope

Figure 4.37 A high-resolution 'structure image' showing a sigma phase precipitate (horizontal) in an aluminium alloy. The direction of viewing is exactly along a ⟨110⟩ of the aluminium matrix and a ⟨100⟩ of the precipitate. No objective aperture was used, so many beams contributed to the image (K. M. Knowles, University of Cambridge)

must also be aligned as accurately as possible, and objective lens astigmatism must be minimized. These factors, combined with the fact that the specimen must be carefully tilted so that the beam direction coincides with a crystallographic axis, demands not only an excellent microscope but great skill and a sophisticated understanding of image formation. Most images cannot be fully understood except by comparison with a series of computed images deduced from likely structures (Figure 4.38). This comparison can be made in a quantitative way by analysing digitized images, i.e. ones in which the image is broken up into individual pixels, which allows the images to be analysed numerically using a computer.

4.3 High voltage electron microscopy (HVEM)

The most obvious effects of increasing the accelerating voltage of a microscope are the reductions in the electron wavelength and in the scattering cross-section. These allow increased penetration (permitting the examination of thicker specimens) and improved resolution (because of equation 1.9). The possibility of removing, to some extent, one of the main limitations of TEM (i.e. the need for a very thin specimen) led to the installation of several 1 MeV microscopes around the world in the 1970s.

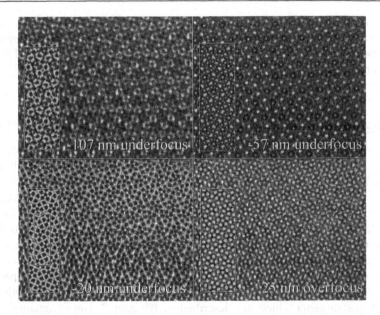

Figure 4.38 High-resolution phase contrast images of a 1·97 nm thick specimen of $Nb_{16}W_{18}O_{94}$, which consists of pentagonal columns of metal atoms arranged in a complex fashion. The unit cell is marked on the −20 nm underfocus image. Note the changes in bright/dark contrast with defocus and the excellent agreement between simulated (boxed) and experimental images. (Courtesy of A. I. Kirkland, University of Cambridge, recorded on the 300 kV FEGTEM installed at Oxford University)

A very high energy electron has a small cross-section for interaction with the electrons of the atoms in the specimen and an even smaller cross-section for interaction with the nucleus. In other words a high voltage electron is very unlikely to cause damage to inter-atomic bonds and extremely unlikely to 'hit' an atom. However, when a 'hit' is scored the potential damage is great. Above a threshold energy it is possible to displace a single atom from its site, and this is the main disadvantage of HVEMs. The fraction of electrons which will cause a displacement is small (say one in ten thousand) but at the beam currents attainable in a TEM (say 10 nA), 10^4 atoms can be displaced per second. The threshold rises with atomic number, but is 170 keV for aluminium and about 400 keV for nickel. This is usually undesirable, but has been exploited in radiation damage experiments, where displacement damage has been used to simulate the effect of neutron irradiation. The greater penetrating power has been invaluable in the study of complex multi-phase alloys, minerals and ceramics. These instruments have also been extensively used for in-situ experiments in which the specimen is heated or cooled, strained or reacted while under observation.

The fact that the cost of manufacturing and running a TEM tends to rise approximately linearly with accelerating voltage, combined with the high displacement damage, has led to the demise of most MeV microscopes. However there are many *medium voltage microscopes*, operating at 300 or 400 keV, which exploit the advantages of higher electron energies.

4.4 Scanning transmission electron microscopy (STEM)

It is possible to combine some of the advantages of the TEM and the scanning electron microscope (SEM) in the single technique called STEM. The beam is focused to as small a spot as possible and is scanned across the specimen while some signal is collected (e.g. X-rays, transmitted, secondary or backscattered electrons). This is discussed briefly in the next chapter (section 5.9.3) but the main reasons for using STEM arise from its potential as an analytical tool. We therefore leave major consideration of the technique until Chapter 7. However, many modern transmission microscopes are equipped with scanning coils which enable them to be used in a STEM mode. Such microscopes are generally called TEM/STEM microscopes in contrast to the 'dedicated STEM' instruments which are described in Chapter 7. Improvements in the electron probe size of TEM/STEM microscopes, particularly those with field emission guns, have now reduced, perhaps even eliminated, the difference in the resolution of the best of the two types of machine.

4.5 Preparation of specimens for TEM

It will be clear from the rest of this chapter that there are many subjects of enormous scientific and technical interest which can be studied using TEM. However, before any observations can be made, a specimen must first be prepared. It is not straightforward to make a specimen thin enough for TEM (a few tens of nanometres to a micron in thickness). The task is made harder still by the need to avoid changes in the specimen due to the preparation technique, and obtain a representative (or sometimes very specific) region. The sample must also be strong enough to handle, and last at least long enough to be examined in the microscope. These are tough requirements, and only rarely are they all met. However it is worth striving for a specimen which gives the maximum amount of information with the minimum amount of complex interpretation, and this means that a great deal of emphasis should be placed on specimen preparation. The techniques employed vary greatly, depending upon the type of material being studied, and form the subject of several books and many scientific papers (e.g. Goodhew 1985). The degree of difficulty in sample preparation can vary from being almost trivial to a skill which can take weeks to master.

The different specimen preparation techniques can be divided into two basic approaches. First is removal of unwanted material, by either chemical or mechanical means, until only a very thin specimen is left behind. Second is cutting, in which the sample is either cut with a knife or cleaved along crystallographic planes so that a very thin specimen, or region of a specimen, is produced. We will describe here only the basics of the most commonly used techniques.

4.5.1 Electropolishing and chemical polishing

The most common technique for thinning electrically conductive materials such as metals and alloys is electropolishing. The principle of the method is that the specimen is made the anode in an electrolytic cell. When a current is passed, metal is dissolved from the anode (the specimen) and deposited on the cathode. The experimental arrangement can be very simple, as shown in Figure 4.39, or can be carried out in a sophisticated semi-automatic commercial unit but the principle is identical. If the composition of the electrolyte and the operating voltage are chosen successfully a specimen in the form of a thin sheet not only becomes thinner but also smoother. Eventually a hole appears in the thin sheet and if the neighbouring regions are sufficiently smooth (i.e. well polished) they will be thin enough for viewing in the TEM. The process is shown schematically in Figure 4.40. Cooling of the electrolyte to low temperatures is often employed to change the kinetics of the etching process to produce a smooth surface. Electropolishing was the technique used to thin the specimens shown in Figures 4.15, 4.17, 4.21 and 4.24.

Automated electropolishers usually take 3 mm diameter disc samples, which emerge with a relatively thick rim supporting the thin central region. These

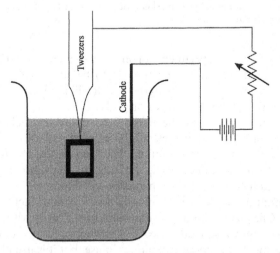

Figure 4.39 Electropolishing as a technique for preparing thin specimens for transmission electron microscopy. At its simplest, a sheet specimen is held in tweezers in a beaker of electrolyte. A small potential (typically 1 to 30 volts) is applied between the specimen and a metal cathode.

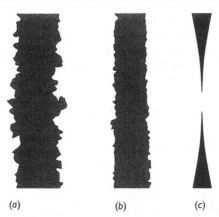

Figure 4.40 The stages of electropolishing a metal specimen. The thick, rough sheet (*a*) becomes smoother and thinner (*b*) and eventually perforates (*c*). The thinnest regions around the perforation should be suitable for examination in the TEM.

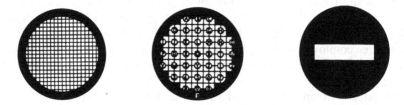

Figure 4.41 Three types of support grid for TEM specimens. From the left are shown a plain mesh grid, a 'finder' grid marked with identification letters and a slot grid for large specimens.

discs fit directly into the specimen holder of the microscope (Figure 4.4(*a*)). Foils prepared by hand will generally need to be supported on a 3 mm grid made of copper or an alternative material which will not interfere with analysis. Figure 4.41 shows a number of types of grid.

There are many variants of the basic electropolishing technique but they share the common feature that they thin a sheet of metal from about 0·1 mm thick to perhaps 0·1 µm in a few minutes. The preparation of a foil from a thin metal sheet is therefore quite rapid, although it may have taken longer to machine or grind it to the starting thickness of 0·1 mm.

A major limitation of electropolishing is that it cannot tackle non-conducting materials. Chemical thinning, using mixtures of acids without an applied potential, is therefore frequently used for ceramics, glasses and semiconductors. There are a wide variety of techniques in use, but it is usually true that the specimen is thinned from one side only. A common technique is to mount a 3 mm disc, usually ground mechanically to a thickness of around 100 µm, onto an inert stub (often PTFE) using a lacquer which is not affected by the etchant.

The stub is either immersed in the etchant or a very fine jet of etchant is directed at the centre of the specimen until a small hole appears at its centre, usually within a few minutes. The finished sample is removed from the stub by dissolving the lacquer in a solvent, and then washed in several changes of solvent to remove all traces of the lacquer. Rotation is often used to produce an even etch across the specimen. The technique is commonly employed for semiconductor specimens, using 1:5 hydrofluoric:nitric acid for silicon and Cl or Br in methanol for most III–V and II–VI type semiconductors.

When the specimen comprises more than one material, for example thin layers deposited on a substrate surface, it is often the case that one or more components of the structure are resistant to the etch used for the bulk of the material. This can be turned to advantage in the case of very thin layers which are electron transparent when the substrate is removed, but more often means that chemical etching of the substrate is followed by ion milling to produce a suitably thin specimen.

Both chemical polishing and electropolishing enjoy the great advantage that they are physically non-damaging. However, this does not mean that they do not have drawbacks, since both techniques can preferentially leach out one or more components of an alloy or compound, causing possible inaccuracies in microanalysis. Also, there seems to be no simple way to predict which etchants or electrolytes will both polish and remove material. For example, it is quite possible to be able to produce a good specimen of one crystalline orientation and a poor one of another orientation using the same etchant. Lists of tried and tested solutions are available, and new recipes are constantly being developed and exchanged.

4.5.2 Mechanical polishing

Most materials science samples will be ground and/or polished as a first stage in the specimen preparation process. Grinding is usually done using paper which has a layer of hard particles, often SiC, stuck onto one face. The paper is graded according to the particle size, which can vary from a sizeable fraction of a millimetre to only a few microns. Usually the paper is placed on a wheel which spins while being lubricated by flowing water. The specimen is mounted on a jig, often using glue or thermoplastic wax, which allows the rate of removal of material to be controlled. The sample is initially ground flat using a coarse grit, and successively finer grits are used to remove the damage inflicted on the sample by the previous stage. Final polishing is often done using diamond powder a micron or less in size, either suspended in oil or water or embedded in a plastic film. Even finer polishing can be performed using mechano-chemical means, often using colloidal silica particles floating in an alkaline liquid.

When one side has been polished, the specimen must then be turned over and the process repeated until the specimen is thin (usually just a few micrometres).

Clearly one must be careful not to thin the specimen away completely, and the final polishing stages are usually performed very carefully. Learning when to change from coarser to finer grinding or polishing steps for a given material can be a painful process; if too fine a grade of paper is used too soon much time will be spent at the grinding wheel, but if it is used too late the specimen will be useless. It is possible to thin some materials to a micron or less in thickness using special polishing pads and jigs (often using a tripod of movable control surfaces to carefully define the polishing plane). By making the specimen in the form of a low angle wedge, it is even possible to make samples which are electron transparent without the need for further treatment. However, most samples which are mechanically thinned need a final ion milling stage before they are suitable for examination (see below). The sample is usually stuck to a grid so that it can be handled using tweezers before it is removed from the jig by melting the wax or dissolving the glue.

There are several variations on this basic process. The technique is widely used for semiconductor cross-section samples, and in this case the region of interest usually lies less than a few microns from the top surface of the sample. This usually means that sacrificial material has to be glued on top of the sample to preserve the device layers. It is also common in such cases that a very specific region has to be examined, and in this case the first grind and polish step has to be performed very carefully so that the region of interest will lie in the electron transparent region of the sample. Tools have also been designed which thin the sample using a small wheel used edge-on and a grinding or polishing slurry, which gives a thick rim around the edge of the sample and added mechanical stability.

4.5.3 Ion and atom milling

If a beam of energetic ions or atoms is directed at a solid, atoms can be knocked out and this process, known as *sputtering*, can be used to thin specimens. A beam of ions or atoms can be generated in a vacuum using *ion* or *atom guns*. There are two types of gun which are commonly used to thin TEM specimens; those using a gas source (usually argon) and field emission ion guns, which use liquid gallium. The latter are generally restricted to focused ion beam microscopes, which are described below. There are a variety of gas source ion and atom guns, but all rely on the ability of a very high electric field (typically 1–5 kV) to generate a plasma in a low pressure gas by stripping one or more electrons from the atoms. The electric field can also be used to accelerate the ions through an aperture in the cathode, producing a beam which is directed onto the TEM sample. Two guns are usually present, so that the sample can be thinned from both sides at once. As in electropolishing, thinning usually proceeds until a small hole in the specimen is produced, which has edges sufficiently thin for TEM investigation.

The sample is usually rotated or oscillated to prevent surface roughness from developing, and if the ions make an angle of less than ten degrees to the surface it is possible to reduce problems due to different ion milling rates in specimens containing several different materials. The rate of removal varies with beam energy, incidence angle and the material being eroded, but is typically a few microns an hour. Many ion milling machines have automatic termination detectors, which switch off the ion guns when a hole appears in the sample, and liquid nitrogen cooling of the sample to reduce the effects of heating from the ion beam. Some degree of control over the ion beam can be obtained using electrostatic lenses, allowing specific areas of a specimen to be thinned.

4.5.4 Focused ion beam (FIB) microscopes

By applying a suitable electric field to molten gallium under vacuum, it is possible to make the liquid droplet form an extremely sharp point. Just as in the case of field emission electron guns, gallium ions can be extracted from the tip – which may be only a few nm wide – and accelerated down an evacuated column using a potential of tens of kilovolts. Since the energy of the ions is very well defined and the source size is very small, such *field emission ion guns* can be used to produce a beam which is only a few nm wide by using an aperture and a demagnifying lens in exactly the same way as a fine electron beam is produced in a scanning electron microscope. Secondary electrons are emitted from the sample where the ions impinge, and so by scanning the ion beam it is possible to obtain an image of the sample, i.e. to make a *focused ion beam (FIB) microscope*. Such fine control of the ion beam allows incredibly precise TEM sections to be manufactured simply by scanning the beam over regions which are to be removed. This is widely used in the semiconductor industry as a failure analysis tool, since, for example, an individual contact in a specific transistor can be sectioned by the focused ion beam microscope and examined using TEM – the only technique which can image structures on the nanometer scale commonly used in state-of-the-art semiconductor devices.

There are of course drawbacks in using a microscope which removes material from the sample while it is being examined. If the region of interest is next to the surface, then it must be protected by a deposited layer to prevent damage from the ion beam. This can be done in the FIB microscope by introducing a metal-organic gas next to the specimen, which decomposes in the ion beam and deposits metal on the sample surface wherever the beam is placed. (The main use of FIB microscopes apart from TEM specimen preparation is 'rewiring' of semiconductor circuits by cutting the existing metal tracks and depositing new ones.) Depositing the metal in the FIB can in itself produce artefacts, since the metal ions are accelerated by the ion beam and can cause amorphization of the sample next to the surface, and so it may be wise to deposit some layer on the surface prior to FIB machining. Another drawback is that the rate of removal of material is relatively slow, which

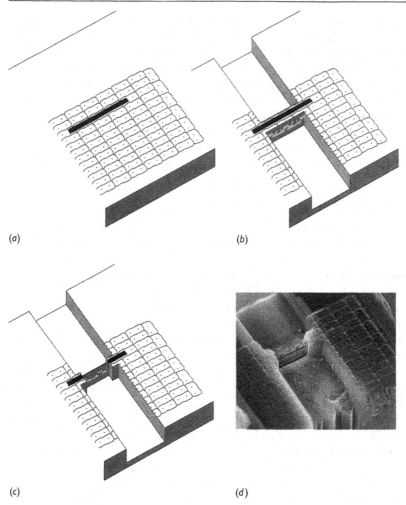

Figure 4.42 The stages in preparation of a FIB-machined TEM cross-section. (*a*) The sample is mechanically thinned to approx. 100 μm and the region of interest identified. A layer of Pt is then deposited over this area (dark) by introducing a metal-organic gas next to the specimen, which decomposes in the focused ion beam. (*b*) Material is removed from either side of the region of interest using a high intensity ion beam. (*c*) Material is progressively removed using a lower intensity ion beam until only a thin membrane is left. (*d*) An actual FIB-machined specimen.

means that samples still require mechanical thinning to a few tens of microns before the final FIB section is made. Figure 4.42 (schematically) shows the different stages in focused ion beam TEM specimen preparation and an example of a FIB/TEM section where a specific region has been thinned to electron transparency.

4.5.5 Cleaving

If the specimen to be examined is a brittle single crystal, it is possible to obtain a small amount of electron transparent material by (carefully) breaking it. Most crystals will prefer to break, or *cleave*, on particular planes. If the material is sufficiently perfect, the surfaces produced by cleaving can be smooth on an atomic level and free of any damage. The symmetry of the crystal usually ensures that two of these planes will meet at an angle of 90 degrees or less, and so when the crystal is cleaved on these planes there will be some thin area along the edge which can be examined in the TEM. The technique is widely used to measure the thickness of epitaxial semiconductor layers; an example is shown in Figure 4.43.

A good cleave is generally obtained by thinning the sample to about 100 µm, and using a fine pointed *diamond scribe* to mark the surface, which acts as the nucleation site for the crack which propagates through the sample. The sample is broken either by pressing a scalpel on its edge, or by turning the sample over and pressing on the opposite surface above the scribe mark using the point of a pair of tweezers. The second cleave is made in a similar manner to the first and the sample is glued onto a grid so that the cleaved edge can be examined in the TEM. The success rate – two perfectly cleaved surfaces – can vary from about 30% when performed manually to close to 100% when using a specially designed cleaving machine. The drawbacks of a cleaved specimen are a very limited thin area and unsuitability for microanalysis (due to the large amount of material close to the thin region, which will give strong X-ray fluorescence), as well as poor control over the position of the thin region. The advantages are very rapid specimen preparation in comparison with mechanical polishing and ion milling – it is possible to make a specimen and examine it in the microscope

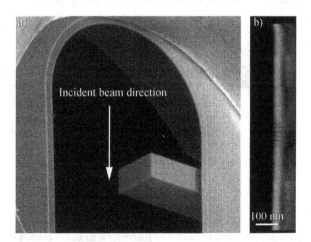

Figure 4.43 (a) An SEM image of a cleaved edge TEM specimen. (b) A dark field 002 image of the specimen in (a) showing the layer structure.

within half an hour – and a very rigid specimen which gives uniform diffraction conditions.

Thin layers deposited on almost any substrate which will cleave rather than tear can be examined using *low angle cleaving*. In this technique, the first cleave is made as usual, but the second cleave is forced to take a particular direction by scribing a line making an angle of only a few degrees to the first. This can produce thin areas a few µm wide, and can be applied to amorphous substrates such as glass. Cleaving under liquid nitrogen can be used to produce brittle fracture in metal layers which are ductile at room temperature.

4.5.6 Ultramicrotomy

An ultramicrotome is a delicate slicing instrument developed from the larger-scale devices used for cutting tissue sections for biological light microscopy. In an ultramicrotome a firmly mounted or embedded specimen with an area less than 1 mm × 1 mm is moved past a fixed knife of glass or diamond. This is shown schematically in Figure 4.44. The resulting slices are collected in a liquid-filled trough and are mounted on grids before being inserted into the microscope.

Ultramicrotomy can be applied to many types of sample in addition to the traditional tissue sections. It is now widely used to prepare polymers for TEM and is increasingly being employed to cut metal sections. Despite the obvious introduction of mechanical damage (such as dislocations) by a cutting technique, chemical information is generally retained. In other words there is minimal redistribution of the elements of the specimen during cutting and analytical microscopy can therefore be successful. For example an ultramicrotome was used to section embedded aluminium alloy powders in order to

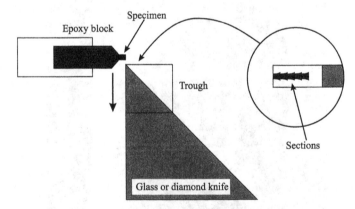

Figure 4.44 The principle of ultramicrotomy. The specimen in its epoxy block is moved steadily past a freshly broken glass edge (the glass 'knife'). The thin sections float in the liquid-filled trough. A diamond can be used as an alternative knife.

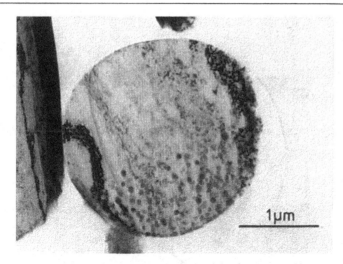

Figure 4.45 A microtomed section of an aluminium alloy particle. The powder was embedded in epoxy resin before sectioning. An extinction contour can be seen at the left of the particle, while small precipitates are visible in the lower third of the specimen.

investigate elemental segregation (or the lack of it) introduced by rapid solidification (Figure 4.45).

Many specimens need to be embedded in resin before sectioning, in order to provide support during cutting. The preparation of a thin section may therefore take a substantial time, although the actual section cutting takes only a few minutes. The operation of an ultramicrotome is highly skilled and needs to be carried out by an expert.

4.5.7 Replication

There is a further method of looking at both biological and non-biological material using the TEM. This is historically a very early method and consists of making a replica of the surface of a specimen rather than trying to thin the whole piece to electron transparency. This can be done surprisingly easily by depositing a thin layer of carbon (or a few other materials) from a source in vacuum. Carbon in an essentially atomic form can be obtained either by striking an arc between two carbon rods or sputtering from a block of graphite. The atomic carbon so produced is deposited on all the surfaces of the vacuum chamber with a line of sight to the source, including the specimen. A very thin layer (only a few tens of nanometres thick) is allowed to build up before the deposition is stopped. The film can be removed from the surface in pieces 1 mm square or larger by floating it onto liquid and then mounted on a copper grid for examination in the microscope.

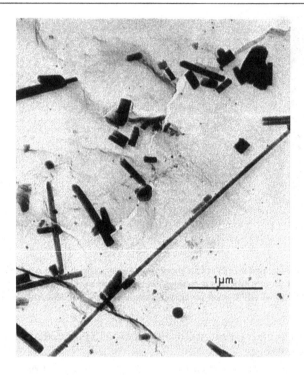

Figure 4.46 An extraction replica showing aluminium nitride particles extracted from a steel (T. N. Baker, Strathclyde University).

The surface structure is revealed either by the different thickness of carbon deposited on the different facets of the surface, or by a second deposition of a heavy metal (e.g. platinum) deposited at a glancing angle to the surface. This technique has been almost completely superseded by scanning probe microscopy (see Chapter 7), which can examine surfaces more quickly, accurately and directly. However, an extension of the technique, the extraction replica, is still widely useful. If the sample contains a finely divided second phase which can be revealed by etching, then particles can be extracted from the sample onto the replica. The extraction thus contains information about the size, shape and distribution of the particles in the original sample. Since the particles are present in the replica they can be studied using electron diffraction or any of the analysis techniques discussed in Chapter 6. An extraction replica from a steel is shown in Figure 4.46.

A very similar procedure to that used for making replicas is used to produce thin carbon films to support fine powders in the TEM. If the carbon is deposited on a glass slide it is almost featureless and can be removed in pieces large enough to cover a 3 mm support grid. Very fine particles can be examined if they are held in the microscope on such a support film.

4.6 Questions

1 List in order of increasing convergence angle: in focus, underfocus, overfocus.
2 What would be the maximum convergence angle for a system with a condenser aperture of 50 mm and a C2 – specimen distance of 20 mm? Express your answer in milliradians.
3 In the two-lens double condenser system what effect does C1 have on the minimum spot size and beam convergence at the specimen?
4 The objective aperture is to be used to allow the undeflected beam to pass but to block electrons diffracted from the (002) planes of copper. What is the maximum size the aperture could be? ($a_{Cu} = 0.360$ nm, $\lambda = 0.0037$ nm, focal length of objective lens = 1 mm).
5 At a magnification of one million times how large is the area of the specimen which you could see at one time on the 100 mm square screen of the TEM?
6 Where would you expect to see the insertion and positioning controls for (a) the objective aperture (b) the selected area aperture (c) the condenser aperture?
7 Describe the difference between bright field and dark field imaging modes.
8 What would you expect to happen to an extinction contour in a TEM image as the specimen is tilted?
9 What would you expect to see in an image of a stacking fault bounded by partial dislocations if: (a) $\mathbf{g} \cdot \mathbf{b} = 1$, $\mathbf{g} \cdot \mathbf{R} = 1$; (b) $\mathbf{g} \cdot \mathbf{b} = 0$, $\mathbf{g} \cdot \mathbf{R} = 1/3$; (c) $\mathbf{g} \cdot \mathbf{b} = 0$, $\mathbf{g} \cdot \mathbf{R} = -2$?
10 What are the three primary contrast mechanisms in TEM?

Chapter 5

The scanning electron microscope

5.1 How it works

The scanning electron microscope (SEM) is similar to the transmission electron microscope (TEM) in that they both employ a beam of electrons directed at the specimen. This means that certain features, such as the electron gun, condenser lenses and vacuum system, are similar in both instruments. However, the ways in which the images are produced and magnified are entirely different, and whereas the TEM provides information about the internal structure of thin specimens, the SEM is primarily used to study the surface, or near surface, structure of bulk specimens. Figure 5.1 is a scanning micrograph showing the surface of a metal alloy. It is much easier for the eye to interpret this type of image than a transmission electron image.

Figure 5.2 is a schematic diagram showing the main components and the mode of operation of a simple SEM, and Figure 5.3 is a photograph of a modern instrument.

The electron source (or gun) is usually of the tungsten filament thermionic emission type, although field emission gun (FEG) sources are increasingly being used for higher resolution (Chapter 2). The electrons are accelerated to an energy which is usually between 1 keV and 30 keV which is considerably lower than the energies typical of the TEM (100–300 keV). Two or three condenser lenses then demagnify the electron beam until, as it hits the specimen, it may have a diameter of only 2–10 nm.

In older instruments, the fine beam of electrons is scanned across the specimen by the scan coils, while a detector counts the number of low energy secondary electrons, or other radiation, given off from each point on the surface. At the same time, the spot of a cathode ray tube (CRT) is scanned across the screen, while the brightness of the spot is modulated by the amplified current from the detector. The electron beam and the CRT spot are both scanned in a similar way to a television receiver, that is, in a rectangular set of straight lines known as a *raster*. In modern instruments, the same effect is achieved by digitally controlling the beam position on the sample and the resultant image is displayed on a computer screen. However, for simplicity

Figure 5.1 Scanning electron micrograph of the surface of a nickel alloy containing dendritic (tree shaped) particles of silica.

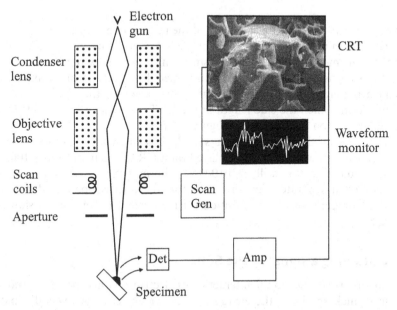

Figure 5.2 Schematic diagram showing the main components of a scanning electron microscope.

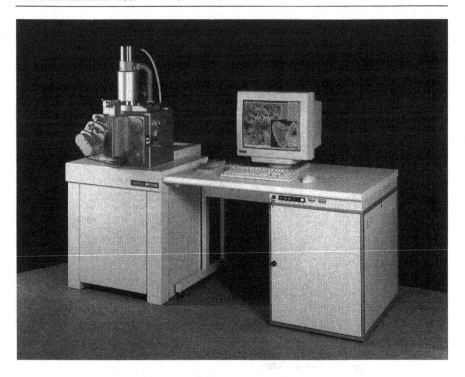

Figure 5.3 A scanning electron microscope (Camscan Ltd).

we will generally discuss the operation of the microscope in terms of a standard analogue scanning system and highlight the differences found in a digital SEM. The mechanism by which the image is magnified is then extremely simple and involves no lenses at all. The raster scanned by the electron beam on the specimen is made smaller than the raster displayed on the CRT. The linear magnification is then the side length of the CRT (L) divided by the side length (l) of the raster on the specimen (Figure 5.4(a)).

For example, if the electron beam is made to scan a raster $10\,\mu m \times 10\,\mu m$ on the specimen, and the image is displayed on a CRT screen $100\,mm \times 100\,mm$, the linear magnification will be $10\,000 \times$. Alternatively, or sometimes simultaneously on a separate waveform monitor, the microscope can display the variation of signal with beam position for the current raster line, as shown in Figure 5.2.

5.2 Obtaining a signal in the SEM

In Chapter 2 we discussed the interaction of electrons with a specimen, and saw that in a thick specimen the energy of the incident electrons was dissipated, resulting in various secondary emissions from the specimen, and that some of

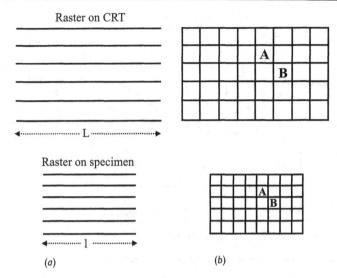

Figure 5.4 (a) The electron beam scans a raster of side *l* on the specimen whilst a raster of side *L* is scanned on the CRT. (b) The rasters can also be thought of as arrays of picture points or pixels.

the inelastically scattered electrons were backscattered out of the specimen. One of the main features of the SEM is that, in principle, any radiation from the specimen or any measurable change in the specimen may be used to provide the signal to modulate the CRT and thus provide contrast in the image. Each signal is the result of some particular interaction between the incident electrons and the specimen, and may provide us with different information about the specimen.

Figure 5.5 shows schematically some of the signals which may be used in the SEM.

All scanning electron microscopes normally have facilities for detecting *secondary electrons* and *backscattered electrons*, and we will discuss their detection below. Of the other radiations, X-rays are used primarily for chemical analysis rather than imaging, and will be discussed in detail in Chapter 6. Auger electrons are of such low energy, and are so easily absorbed that they require an ultra high vacuum system and specialized equipment for their efficient use. Auger spectroscopy and the scanning Auger microscope are important surface analytical techniques which will be discussed in Chapter 7. The other signals have important but more specialized applications, and we will discuss them later in this chapter.

We discussed the trajectory of an electron through a solid in Chapter 2. Monte Carlo simulations such as that shown in Figure 2.5, and also direct experiments, have shown that the electrons are scattered in the specimen within a region such as that shown in Figure 5.6.

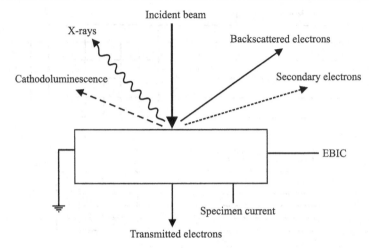

Figure 5.5 Some of the signals which may be used in the SEM.

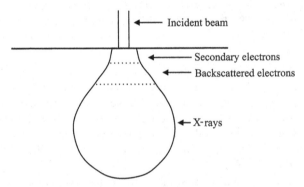

Figure 5.6 The interaction volume and the regions from which secondary electrons, backscattered electrons, and X-rays may be detected.

The region into which the electrons penetrate the specimen is known as the interaction volume, and throughout it, the various radiations are generated as a result of inelastic scattering, although as the primary electrons lose energy the amount and type of the secondary radiations will alter.

Even though radiation is generated within this volume, it will not be detected unless it escapes from the specimen, and this will depend on the radiation and the specimen. Thus, X-rays are not easily absorbed, and most will escape from the specimen. Therefore, as will be discussed in more detail in Chapter 6, the volume of material contributing to the X-ray signal, or *sampling volume*, is of the same order as the interaction volume, which may be several micrometres in diameter. Electrons will not be backscattered out of the specimen if they have penetrated more than a fraction of a micrometre, and the backscattered signal

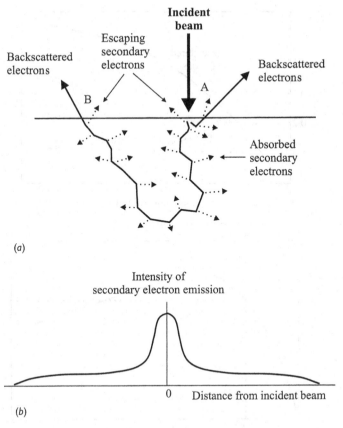

(a)

(b)

Figure 5.7 (a) The generation of secondary electrons. (b) The spatial distribution of secondary electrons.

therefore originates from a much smaller region as shown in Figure 5.6. From Figure 2.11 it can be seen that the backscattered electrons have a broad energy spread. Those of the highest energy are electrons which have been scattered only a few times. These originate near the incident beam (A in Figure 5.7(a)) and are capable of giving information at high spatial resolution as well as yielding crystallographic information (see section 5.8). Electrons which have undergone multiple scattering (B in Figure 5.7(a)) lose more energy, come from a larger area and therefore yield information at worse spatial resolution.

Although secondary electrons are generated both by the primary electrons entering the specimen and by the escaping backscattered electrons, the former are more numerous, and therefore the detected secondary electron signal originates mainly from a region which is little larger than the diameter of the incident beam, as shown schematically in Figure 5.7. When we come to discuss the resolution of the SEM, we will find that it is closely related to the sampling

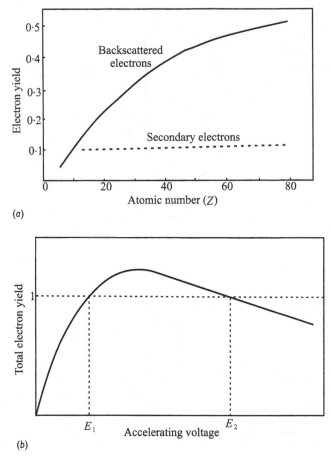

Figure 5.8 (a) The effect of atomic number on the yield of backscattered electrons (η) (after K. F. J. Heinrich), and secondary electrons (δ) (after D. B. Wittry). (b) The effect of accelerating voltage on the total electron yield ($\eta + \delta$).

volume of the signal used. Secondary electrons, having the smallest sampling volume, are therefore capable of giving a better spatial resolution than the other signals which we have discussed.

The numbers of secondary and backscattered electrons emitted from the specimen for each incident electron are known as the *secondary electron coefficient* (δ) and the *backscattered electron coefficient* (η) respectively.

As shown in Figure 5.8(*a*), η is strongly dependent on the atomic number of the specimen whereas δ is not. As shown in Figure 5.8(*b*), the total electron yield is a complex function of the accelerating voltage, with a maximum between 1 and 5 keV. As η is almost independent of voltage, Figure 5.8(*b*) effectively represents the voltage dependence of δ.

If the electron yield is not equal to unity, then unless the sample is a conductor, it will tend to become charged during examination as discussed in section 5.11. Figure 5.8(*b*) indicates that careful control of the accelerating voltage can produce an electron yield of unity which will overcome this effect (see section 5.11).

5.2.1 Detecting secondary electrons

By far the most widely used signal in the scanning electron microscope is that from secondary electrons. Secondary electrons are detected by a scintillator–photomultiplier system known as the Everhart–Thornley detector. The detector is shown schematically in Figure 5.9.

The secondary electrons strike a scintillator, e.g. a phosphor, which then emits light. The light is transmitted through a light pipe, and into a photomultiplier which converts the photons into pulses of electrons, which may then be amplified and used to modulate the intensity of the CRT.

The energy of the secondary electrons (10–50 eV) is too low to excite the scintillator, and so they are first accelerated by applying a bias voltage of $\sim +10\,keV$ to a thin aluminium film covering the scintillator. A metal grid or collector, at a potential of several hundred volts, surrounds the scintillator and this has two purposes. First it prevents the high voltage of the scintillator affecting the incident electron beam, and secondly, it improves the collection efficiency by attracting secondary electrons, and thus collecting even those which were initially not moving towards the detector, as shown in

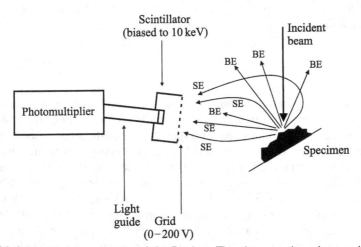

Figure 5.9 Schematic representation of the Everhart–Thornley secondary electron detector, showing the paths of secondary (SE) and backscattered (BE) electrons.

Figure 5.9. The Everhart–Thornley detector system is very efficient, and for flat specimens, almost all the secondary electrons are collected.

5.2.2 Detecting backscattered electrons

Backscattered electrons which are travelling in the appropriate direction will of course hit the scintillator of the Everhart–Thornley detector and be detected. Therefore we should note that the signal discussed in the previous section is not purely due to secondary electrons as it does contain a small backscattered component. If the scintillator bias is switched off, or the collector given a small negative voltage, then secondaries are excluded from the detector, and the backscattered signal is obtained. However, only those electrons travelling along the direct line of sight towards the detector will be collected, and thus the geometric efficiency is very low. This method of detection is now rarely used, and most microscopes are fitted with purpose built backscattered electron detectors, which can be of three types.

Scintillator detectors

These detectors are of the scintillator/light pipe/photomultiplier type, and are designed to maximize the solid angle of collection. A good example of this is the Robinson detector shown schematically in Figure 5.10(a). The advantage of these detectors is their rapid response time, which means that, like the Everhart–Thornley detector, they may be used in conjunction with rapid scan rates. However, they are bulky, and may restrict the working distance of the microscope, and may need to be retracted if, for example, it is necessary to detect X-rays.

Solid-state detectors

When a high-energy electron impinges on a semiconductor, it produces many electron-hole pairs. Normally these will rapidly recombine, but if a voltage is applied to the semiconductor, for example, by the self-bias generated by a P–N junction, then they may be separated, thus producing a current, which can subsequently be amplified. Figure 5.10(b) is a schematic diagram of such a detector. The detector is usually in the form of a thin flat plate, which is mounted on the objective polepiece, and thus does not interfere with normal operation of the instrument. The detector consists of up to four such elements whose outputs may be measured independently. The main disadvantage of a solid-state detector is its relatively slow response time, and hence its unsuitability for rapid scan rates.

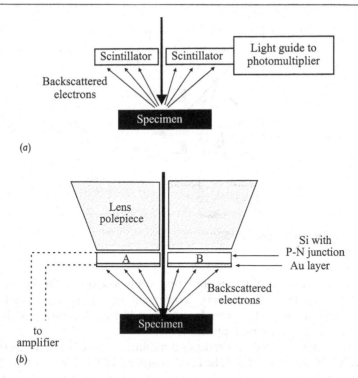

Figure 5.10 (a) A large area, Robinson type scintillator detector. (b) A solid state silicon detector with two elements A and B.

Through-the-lens detectors

In high resolution SEMs a different type of electron detection system is sometimes used. Such instruments have specially designed objective lenses with very large magnetic fields and low spherical aberration and the sample may be placed within the strong magnetic field of the lens. A scintillator detector is placed within the lens and backscattered and secondary electrons pass upwards through the lens to this detector. Such a system has a very good collection efficiency and allows the microscope to operate at very short working distances. However in some instruments such an arrangement places severe restrictions on the size and movement of the sample.

5.3 The optics of the SEM

The purpose of the lenses in the SEM is to produce a fine beam of electrons incident on the specimen. Figure 5.11 is a simplified ray diagram of a microscope which has two lenses, a condenser lens and an objective lens. We can understand the main features of the microscope by treating the electromagnetic

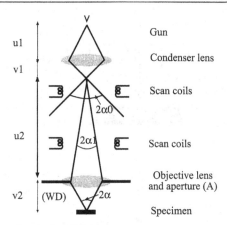

Figure 5.11 Ray diagram of a two-lens SEM.

lenses as thin convex optical lenses, and using geometric optics theory. For the moment, we will ignore lens defects. You might find it useful to study the MATTER software module, 'Introduction to Electron Microscopes' if you are unfamiliar with geometrical optics.

The electron gun (Chapter 2) produces a monochromatic beam of electrons, with a current in the beam of I_0. The condenser lens, of focal length f_c, collects most of these electrons and produces a demagnified image of the filament at a distance v_1 from the condenser lens. If the diameter of the filament is d_0 then the diameter of the intermediate filament image (d_1) is given by

$$d_1 = d_0 \times v_1/u_1 \tag{5.1}$$

and v_1 is obtained from the thin lens formula, equation 1.1. The objective lens of focal length f_0 is used to further demagnify the filament image, producing a probe of diameter d on the surface of the specimen, which is a distance v_2 below the objective lens. The distance v_2 is known as the working distance (WD) of the microscope.

The diameter of the final probe on the specimen is then

$$d = d_1 \times v_2/u_2 = d_1 \times WD/u_2 \tag{5.2}$$

It can be seen that if the strength of the condenser lens is increased, v_1 decreases, and the intermediate beam diameter d_1 decreases. Also u_2 must increase, as $u_2 + v_1$ is constant, and therefore the demagnification of the objective also increases, and the probe diameter decreases. The probe size in the SEM is therefore regulated by altering the strength of the condenser lens. For a constant condenser lens setting, equation 5.2 shows that the probe diameter also decreases as the working distance is decreased.

In order to minimize spherical aberration (Chapter 2), the entry of rays into the objective lens is restricted by an aperture of diameter A, and therefore, as

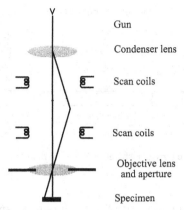

Figure 5.12 Scanning of the electron beam by two sets of coils so that the beam always passes
.... through the optic axis of the objective lens.

may be seen from Figure 5.11, not all of the electron beam which passes
through the condenser lens can enter the objective lens. If the semi-angle of
the rays leaving the condenser lens is α_0, and the semi-angle of the rays entering
the objective lens is α_1, then the current in the final probe is

$$I_1 = I_0 \times (\alpha_1/\alpha_0)^2. \tag{5.3}$$

The current therefore decreases as the condenser lens strength increases (i.e. as
the probe becomes smaller), and also decreases as the aperture diameter (A) is
reduced.

When we come to consider the ultimate resolution of the SEM, we will find
that the probe size, and the current in the probe, play important roles in
determining the performance of the microscope.

Deflection of the beam is accomplished by energizing a pair of coils as shown
schematically in Figure 5.12. In order to scan a raster, two orthogonal pairs of
coils are required. The SEM normally has two such sets of coils, which are set
to deflect the beam in opposite directions. Thus, as seen in Figure 5.12, the
beam scans across the specimen, but always passes through the optic axis at the
objective lens.

5.4 The performance of the SEM

5.4.1 Pixels

The process of image formation in the scanning electron microscope is quite
unlike the formation of an image in the optical or transmission electron microscope, as the image is built up sequentially during the scan. A very useful
concept in understanding the imaging performance of the SEM is that of the
picture element or *pixel*. In an analogue SEM, the amplified signal from the

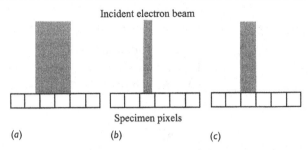

Incident electron beam

Specimen pixels

(a)   (b)   (c)

Figure 5.13 The relationship between the electron beam diameter and the size of the specimen pixels.

detector is output to a high quality cathode ray tube, and the minimum size of spot which may be obtained on such a CRT is typically ~ 0.1 mm (100 μm). A 100 mm square CRT, such as is used for photographic image recording can therefore contain 1000×1000 discrete picture elements or *pixels*. In a digital SEM the image will be recorded on a framestore which may contain a similar number of pixels. As shown in Figure 5.4, the spot on the CRT (or the framestore pixel) mimics the movement of the electron beam on the specimen, and therefore for each of the pixels on the CRT there is a corresponding pixel on the specimen (Figure 5.4(b)). The size of the specimen pixel (p) is given by

$$p = \frac{100}{M} \mu m \qquad (5.4)$$

where M is the magnification.

Now the resolution was defined in Chapter 1 as the smallest separation of two points that the microscope can detect as separate entities. It is clear that in order to resolve two features, e.g. A and B in Figure 5.4(b), they must occupy separate pixels. Therefore, the working resolution of the instrument can be no better than the specimen pixel size p as given by equation 5.4.

The size of the electron probe relative to the specimen pixel size is very important.

If the electron probe is larger than the specimen pixel (Figure 5.13(a)), then the signal from adjacent pixels is merged, and the resolution is degraded. If the electron probe is smaller than the specimen pixel (Figure 5.13 (b)) then the signal will be weaker, and, as we will discuss later, may be noisy. It is clear that for optimum performance of the instrument, we should in general aim to make the probe diameter (or more correctly the *sampling volume* as discussed in section 5.2) equal to the specimen pixel diameter (Figure 5.13(c)). This means that for optimum performance, the probe size should be adjusted as the magnification of the microscope is altered.

5.4.2 Depth of field

Apart from its good spatial resolution, one of the most important aspects of the scanning electron microscope is its large depth of field. If we compare images taken by an optical microscope and an SEM of the same object, at similar magnifications, as in Figure 5.14, then the difference is striking.

The important consequences of the large depth of field of the SEM cannot be overemphasized. As an example, over the past thirty years the SEM has contributed greatly to our understanding of fracture processes by providing a means of examining fracture surfaces at high resolution.

Figure 1.11 shows the electron beam emerging from the objective aperture (diameter A) and incident on a specimen. Although the beam is focused on the specimen, the convergence angle α is small, and, assuming a point focus, the beam diameter defocuses by less than s over a vertical distance of h, where

$$s = h\alpha \tag{5.5}$$

If the defocus is no greater than a specimen pixel, then the image will remain in focus, and thus from equations 5.4 and 5.5, the distance h over which the specimen will remain in focus, the depth of field, is given by

$$h = \frac{0.1}{M\alpha} \text{mm} \tag{5.6}$$

From Figure 5.11, we see that the convergence angle α of the beam is given by

$$\alpha = \frac{A}{2WD} \tag{5.7}$$

Thus from equations 5.6 and 5.7, we find that the depth of field is

$$h = \frac{0.2WD}{AM} \text{mm} \tag{5.8}$$

For typical values of aperture diameter of 100 μm, and a working distance of 20 mm, equation 5.8 shows that at a magnification of 1000×, the depth of field is 40 μm. By comparison, an optical microscope working at a magnification of 1000× with an objective lens of numerical aperture of 0.7 would have a depth of field of only ~1 μm. You can experiment with these variables in the MATTER module, 'Using the Scanning Electron Microscope'.

5.5 The ultimate resolution of the SEM

We saw in the last section that the best working resolution of the SEM was the specimen pixel size p, given by equation 5.4, and that this depended on the magnification of the instrument. However, this resolution is only achieved if the diameter of the beam sampling volume is no larger than p. As discussed in section 5.2, and shown in Figure 5.6, the sampling volume depends on the signal being used. For example X-rays have large sampling volumes which

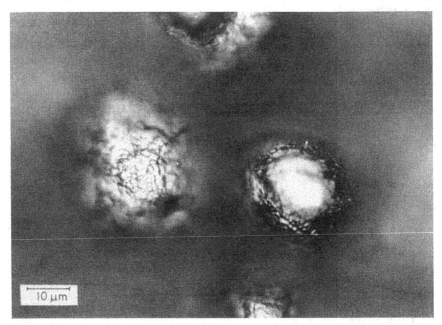

(a)

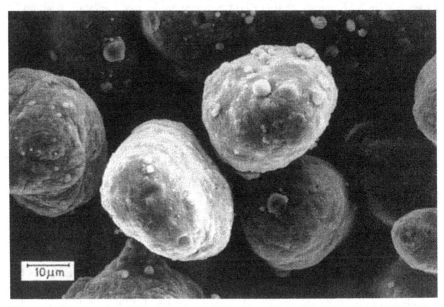

(b)

Figure 5.14 Particles of rapidly solidified aluminium alloy powder. (a) Imaged in the optical microscope (numerical aperture = 0·65). (b) Seen in the scanning electron microscope.

are comparable with the total interaction volume, and the diameter of the sampling volume in this case is several micrometres. No matter how good our microscope is, or how well we adjust it, we cannot achieve a spatial resolution which is better than this. Secondary electrons have the smallest sampling volume, with a diameter little larger than the probe diameter.

However, as the probe diameter is reduced, the beam current is decreased, as discussed in section 5.4, and ultimately the beam current will be insufficient to generate a usable signal. Thus we can define the ultimate resolution of the SEM as being that of *the smallest probe which can provide an adequate signal from the specimen.*

5.5.1 The minimum attainable probe size

As discussed in section 5.3, the probe size may be decreased by increasing the strength of the condenser lens and decreasing the working distance. However, the limitations on focusing the electron beam to a small spot with a lens are identical to those which limit the resolution of images in the TEM (chapter 1 and equations 1.7–1.9). As the working distance decreases the beam convergence angle α increases (Figure 5.11). Rays which are off the optic axis are subject to spherical aberration and instead of a point focus on the specimen, we obtain a disc of diameter d_s, where

$$d_s = 2C_s\alpha^3 \tag{5.9}$$

where C_s is the coefficient of spherical aberration of the lens.

There is also aberration introduced by diffraction at the aperture (Chapter 1), which, for electrons of wavelength λ, limits the minimum spot size to

$$d_d = 1 \cdot 22\lambda/\alpha \tag{5.10}$$

If the microscope is adjusted so as to give a probe of the theoretical diameter d_1, according to equation 5.2, then this value will be increased by the aberrations d_s and d_d. The real probe size d may then be taken as

$$d = (d_1^2 + d_s^2 + d_d^2)^{1/2} \tag{5.11}$$

The minimum value of d, corresponding to the case when $d_1 = 0$, is obtained by minimizing the aberrations, and is given by

$$d_{\min} = K\lambda^{3/4} C_s^{1/4} \tag{5.12}$$

where the constant K is about unity.

For a typical SEM, operating at 20 keV, with an objective lens of $C_s = 20$ mm, $d_{\min}$ is 2·3 nm, and $\alpha = 5\cdot1 \times 10^{-3}$ radians.

As discussed in section 5.3, as the probe diameter (d) is decreased, so the current (I) in the beam decreases. The relationship between these parameters is given by Pease and Nixon, for a thermionic emission filament, as

$$d = d_{min} \left[7 \cdot 92 \times 10^9 \left(\frac{IT}{j} \right) + 1 \right]^{3/8} \tag{5.13}$$

where T is the filament temperature and j is the current density at the filament surface.

5.5.2 The minimum usable beam current

In order to resolve two points on the specimen, there must be a discernible difference between the signals from these two regions.

Figure 5.15(a) shows how the intrinsic signal from the specimen might vary with scan position. If we compare the signal S_{max} from one point of the specimen, with the signal S from an adjacent point, then the contrast from the specimen C, which lies between 0 and 1, and is sometimes called the *natural contrast*, is defined as

$$C = \frac{(S_{max} - S)}{S_{max}} = \frac{\Delta S}{S_{max}} \tag{5.14}$$

Now the signal that is detected in the SEM is not a continuous signal, but for each pixel is derived from the number of secondary electrons n arriving at the detector in a fixed time period. Because these events are randomly distributed in time, simple statistical theory tells us that if the average number of electrons detected from a particular point on the specimen is $\bar{n}$, then $\bar{n}$ will vary by an amount up to $\sqrt{\bar{n}}$ about the mean. The noise N is then defined as $\sqrt{\bar{n}}$.

In a real situation therefore, the variation of signal with scan position will be as shown in Figure 5.15(b), with the noise tending to obscure the natural contrast of the specimen. Whether or not the observer can detect the two points of interest, i.e. whether they can see the contrast, is a physiological problem.

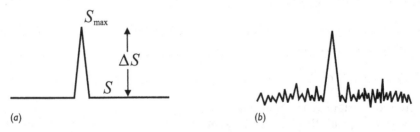

(a) (b)

Figure 5.15 (a) Idealized signal waveform showing contrast. (b) A similar signal with noise.

Rose has determined that the human eye can only distinguish two points on a CRT if

$$\Delta S > 5N \quad \text{or} \quad \Delta S > 5\sqrt{\bar{n}} \tag{5.15}$$

We can use this criterion to determine the minimum level of contrast which can be observed. You can experiment with your own perception of contrast in the MATTER module 'Image Fundamentals'. Combining equations 5.14 and 5.15, we find that the minimum level of contrast which can be observed is given by

$$C > \frac{5}{\sqrt{\bar{n}}} \tag{5.16}$$

The minimum level of signal necessary for us to observe a contrast level of C in the specimen is

$$\bar{n} > (5/C)^2 \tag{5.17}$$

We can relate $\bar{n}$, the mean number of electrons detected for each pixel to the operating conditions of a microscope with a beam current I and a frame scan time F.

If we assume, as before, that we have a CRT or framestore of 1000×1000 pixel resolution (total of 10^6 pixels), then the time t that the beam dwells on a particular pixel is $F \times 10^{-6}$. The number of electrons (of charge e) which enter the specimen in this time is therefore

$$n_0 = \frac{It}{e} = \frac{IF \times 10^{-6}}{e} \tag{5.18}$$

The number of electrons actually detected n will depend on the beam–specimen interaction (Chapter 2), and the efficiency of the detector. We can write

$$n = qn_0 \tag{5.19}$$

where q is the product of the detector efficiency and the electron yield. For secondary electrons, the former is approximately unity, and the latter is ~ 0.1–0.2, giving a value for q of between 0.1 and 0.2.

By combining equations 5.17, 5.18 and 5.19, and putting $e = 1.6 \times 10^{-19}$ C, we can now express the Rose criterion in terms of the critical current I_c, which is required to discern a contrast level C in the specimen as

$$I_c > \frac{4 \times 10^{-12}}{qFC^2} \text{ amperes} \tag{5.20}$$

We can see from this equation that for a given detection system, there is a minimum beam current required to observe a particular contrast level, and that this current increases as the frame scan time decreases.

If we substitute I_c obtained from equation 5.20 into equation 5.13, then we can predict the minimum probe size, and therefore the best resolution obtainable in terms of a given contrast level in the specimen. This is shown

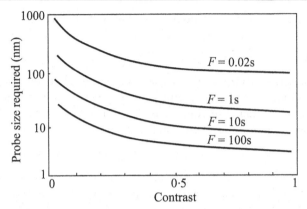

Figure 5.16 The minimum size of probe which may be used to observe a given level of contrast in the signal, as a function of the frame scan time. The microscope parameters for which these relationships hold are discussed in the text.

graphically in Figure 5.16, for a typical SEM using secondary electrons. The effect of frame scan time F is also shown, for F varying between 100 s, which is a realistic value for image recording (see section 5.10) and 0·02 s which is the television scan rate.

In constructing this graph, the following parameters have been assumed:

$$q = 0.2, \; j = 4\text{A}/\text{cm}^2, \; T = 2800\,\text{K}, \; C_s = 20\,\text{mm}, \; E = 20\,\text{keV}$$

Although electron microscope manufacturers will specify the ultimate resolution of an instrument, it is important to realize that this is for a contrast value of unity, which will only be obtained in an ideal specimen. As may be seen from Figure 5.16, for specimens of low natural contrast the ultimate resolution may be substantially lower.

5.5.3 High-performance microscopes

In discussing the performance of the scanning electron microscope so far, we have based our calculations on the parameters and operating conditions of a typical SEM with a tungsten filament and an objective lens C_s of 20 mm, which will have an ultimate resolution of around 5 nm. The discussions above have shown that the resolution of the SEM is limited by the spherical aberration of the objective lens and by the beam current, and substantially better resolution than this can be achieved by improving one or both of these parameters.

The coefficient of spherical aberration of a good TEM objective lens is typically 2 mm. By putting scan coils into a TEM, then extremely good spatial resolution can be achieved, albeit at the cost of limiting the specimen size. The resolution of such a TEM/SEM instrument is of the order of 2 nm.

An alternative is to use a higher brightness electron gun and LaB$_6$ filaments (ten times brighter) or field emission sources (100 times brighter) may be used. Field emission gun microscopes (FEGSEM) which have spatial resolutions approaching ~1 nm are now readily available and such systems have a number of other advantages including a much improved performance at low accelerating voltages.

5.6 Topographic images

One of the principal uses of the scanning electron microscope is to study the surface features, or *topography*, of a sample. The discussion in the previous sections has shown that in order to obtain an image in the microscope, we must have some variation in the signal obtained from different parts of the specimen.

Although topographic images may be obtained using most signals, we will only consider the use of secondary and backscattered electrons. The backscattered electron coefficient η and the secondary electron coefficient δ, are both a minimum when the surface of the specimen is perpendicular to the electron beam. This is because of the shape of the interaction volume and its relationship to the surface of the specimen, as shown in Figure 5.17. As the specimen is tilted, electrons are increasingly likely to be scattered out of the specimen, rather than further into the specimen.

The secondary electron coefficient varies with the tilt of the specimen θ, as $\delta = \delta_0 \sec \theta$. Therefore, more secondaries are produced from tilted regions of the specimen. As the efficiency of the Everhart–Thornley detector is not very sensitive to the trajectories of the secondary electrons, we expect the number of detected electrons therefore to increase with surface tilt. It is partly for this reason that specimens being studied for topographic contrast detector are usually tilted some 20–40 degrees towards the detector (Figure 5.9).

Topographic images obtained with secondary electrons, such as Figure 5.1, look remarkably like images of solid objects viewed with light. As we are accustomed to this sort of image, we find these topographic images easy to interpret. This similarity arises because of the very strong analogy between the two imaging modes. Figure 5.18(*a*) shows a schematic diagram of a surface under diffuse lighting conditions.

Light is arriving from almost all directions so that whatever the orientation of each of the facets A, B and C some light is reflected towards the eye. Figure 5.18(*c*) shows the analogous situation for secondary electrons in the SEM. The only major difference is that the electrons are travelling in the opposite direction to the light rays. The specimen in the SEM therefore appears as if we were looking at it from above, when it is being illuminated with diffuse light, the detector being the source of the diffuse light.

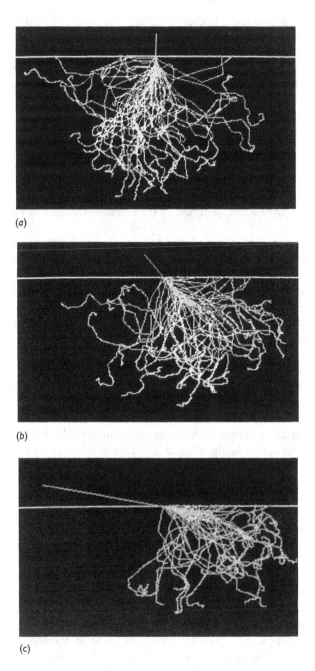

Figure 5.17 The effect of beam tilt on the interaction volume and on the backscattering coefficient (η) of aluminium at 20 keV. The figures were produced using the SEM
module of the MATTER software suite. (*a*) Tilt = 0°, η = 0.13. (*b*) Tilt = 40°,
η = 0.21. (c) Tilt = 80°, η = 0.55.

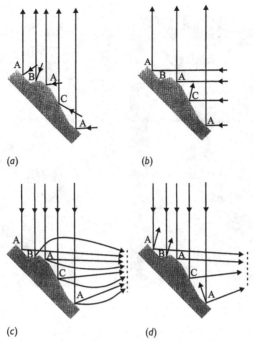

(a)   (b)

(c)   (d)

Figure 5.18 The analogy between the eye and the SEM. The four diagrams show the same specimen viewed under different conditions. (a) Diffuse illumination viewed by the eye
(looking down from the top of the page). (b) Direct illumination viewed by the
eye. (c) The equivalent of (a) in the SEM (secondary electrons). (d) The equivalent
of (b) in the SEM (backscattered electrons).

Figure 5.19(a), which shows a faceted fracture surface viewed in the SEM
with secondary electrons, is typical of the topographic images of interest to
materials scientists.

It is possible to obtain even more information about the specimen if
stereomicroscopy is used. This is achieved by taking two pictures under
identical conditions, except that between exposures the specimen is tilted
by a few degrees. These then represent the views which our left and right
eyes would see. If we view the pictures so that the left eye sees the left image,
and the right eye the right image, our brain will interpret the combined
images as a single three-dimensional picture. The easiest way of achieving
this is to mount the photographs side by side, so that the same feature is
separated by about 6 cm in the two photographs, and to use a special stereo
viewer. However, with a little practice, most people can train themselves to
get the same effect by simply relaxing the eyes so that they focus on infinity,
and viewing the images, concentrating on a single prominent feature, whose
two images will (hopefully) eventually combine. Figure 5.20 is a stereo pair of
a corroded metal surface.

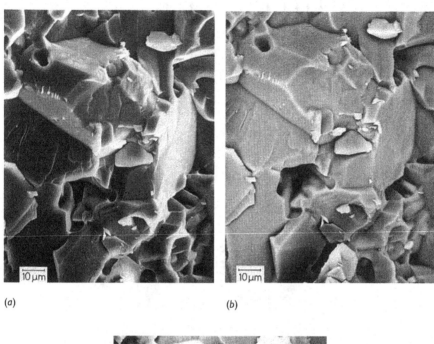

(a)

(b)

(c)

Figure 5.19 Scanning electron micrographs of the fracture surface of alumina. (a) Secondary electron image. (b) Same area, imaged with all four segments of a solid state backscattered electron detector. (c) Same area, using only a single segment of the detector.

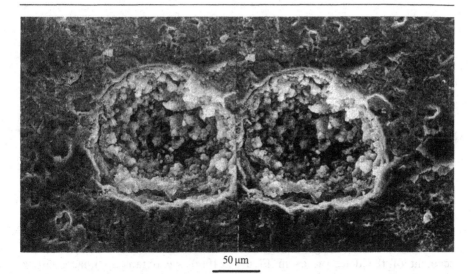

50 μm

Figure 5.20 A stereo pair micrograph of a corrosion pit in a powerstation condenser tube. The three-dimensional effect can be observed by viewing the pair of micrographs using a special viewer, or, with practice, using the unaided eye. (G. Gibbs and J. E. Castle, University of Surrey)

The optimum amount of tilt between the two images is typically 5–10 degrees, but this will depend on the magnification and the height differences in the specimen. It is possible to obtain quantitative measurements from stereo pairs by determining the relative lateral displacement or parallax p of different regions of the image. The vertical height separation h, of two features in the image is given by

$$h = p/2M \sin \theta \qquad (5.21)$$

where M = magnification and θ = half the tilt angle.

Measurements may be made directly on the micrographs, or by using a stereometer such as is used for aerial surveying.

Stereoscopic images may be viewed directly in the microscope in some instruments. This may be achieved for example by rapidly tilting the sample between two positions, displaying the two images in red and green and viewing them with appropriately coloured glasses (anaglyph technique).

Topographic images may also be obtained using backscattered electrons. The yield of backscattered electrons also increases with increasing tilt angle, but the direction of the electrons becomes progressively more peaked in the forward direction as seen in Figure 5.17. As only 'line of sight' backscattered electrons are detected, this means that the number of electrons detected is a function of both the specimen and the detector geometries. Once again there is a close analogy with viewing a specimen by eye. Figure 5.18(b) shows a surface illuminated by a parallel beam of light such as a spotlamp.

Only the A facets are now correctly oriented to reflect light to the eye. Facet C reflects the light away from the eye, while facet B cannot even be reached by the illuminating beam. The analogous situation for a small backscattered electron detector in the SEM is shown in Figure 5.18(d). The specimen in the SEM appears as if we were looking at it from above, when it is illuminated by a light source making the same solid angle with the specimen as does the detector. Therefore in the SEM, a rough surface observed with backscattered electrons will have more shadows than the same specimen viewed with secondary electrons. Viewing a specimen with a multi-element backscattered detector (Figure 5.10(b)), is similar to viewing a specimen with a battery of spotlights. If we use all the (four) segments, then, as shown in Figure 5.19(b), we obtain an image which has a few more shadows and highlights than the secondary electron image of Figure 5.19(a). However if we only use one segment of the detector, as in Figure 5.19(c), we obtain an image with a much harsher appearance. In some circumstances, it may be easier to interpret such an image, and to distinguish the peaks from the troughs in the specimen, than for the secondary electron image.

A multi-element backscattered electron detector may also be used to enhance the topographic image from a specimen which is almost flat. The elements A and B in Figure 5.10(b), being on opposite sides of the optic axis, will clearly receive different topographic signals from the specimen. If the signals from these elements are *subtracted* from each other, then we will obtain no contrast from flat regions of the specimen, but contrast from any topographic features will be enhanced. This mode of operation is particularly useful when there are other sources of contrast in the specimen, such as compositional or crystallographic contrast, as these are then suppressed. An example of this effect is seen in Figure 5.21(a).

Although backscattered electrons give us a very versatile approach to topographic imaging, we must remember that the sampling volume for the backscattered electrons is large, and resolution may be limited to $\sim 0.1\,\mu\text{m}$.

5.7 Compositional images

The signal from the specimen is capable of yielding information not only about the surface topography, but also about the composition. The secondary electron coefficient does not depend very much on the composition of the sample, although it may be sensitive to the surface condition and electronic structure of the material. However, the backscattered coefficient η varies monotonically with atomic number Z as shown in Figure 5.8(a).

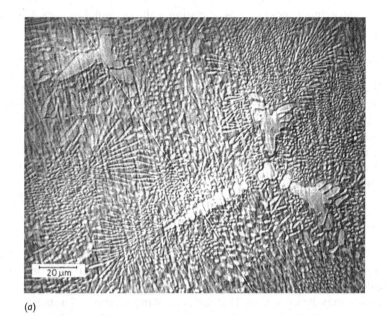

(a)

(b)

Figure 5.21 Backscattered images of a mechanically polished silver soldered joint. (*a*) Topographic contrast arising from the different resistance to abrasion of the phases is enhanced by subtracting the signals from the two elements of the detector. (*b*) By adding the signals from the two detector elements, topographic contrast is suppressed, and compositional contrast is enhanced.

η is almost independent of accelerating voltage, and Heinrich has shown that it is given by

$$\eta = -0.254 + 0.016Z - 1.86 \times 10^{-4}Z^2 + 8.3 \times 10^{-7}Z^3 \qquad (5.22)$$

The backscattered coefficient from a compound or phase containing several elements may generally be obtained from equation 5.22 by using a rule of mixtures based on weight fraction, although there may be a small dependence of η on density.

The magnitude of the compositional or *atomic number contrast* from two phases of backscattered coefficients η_1 and η_2, where $\eta_1 > \eta_2$ is readily calculated using equation 5.14 as

$$C = \frac{\eta_1 - \eta_2}{\eta_1} \qquad (5.23)$$

Table 5.1 gives some examples of atomic number contrast calculated from equation 5.23. It may be seen from Table 5.1 that the contrast from adjacent elements is quite small, typically 1–5%, and that contrast from different phases in an alloy may be even less. Therefore, in many systems of interest, this is a relatively weak form of contrast compared to topographic contrast which may approach 100 %. Specimens for compositional imaging should therefore preferably be polished flat. Unetched, metallographically polished specimens are ideal for such examination, and an example is shown in Figure 5.21. It is instructive to compare Figures 5.21(*a*) and (*b*) which were taken under identical conditions. For Figure 5.21(*a*), as discussed above, the detector signals from elements on opposite sides of the polepiece were subtracted. This removes the compositional contrast, but reveals the weak topographical contrast from the polishing scratches and hardness differences of the phases. Figure 5.21(*b*) was obtained by adding the signals from the different detector elements. This minimizes the topographic contrast, and enhances the atomic number contrast which depends only on the number of backscattered electrons collected by the total detector. For compositional imaging, the solid angle of the detector should be as large as possible, and therefore short working

Table 5.1 Atomic number contrast

Phase 1	Z_1	Phase 2	Z_2	η_1	η_2	Contrast %	Resolution Degradation (nm)
Al	13	Mg	12	0·153	0·141	7·6	19
Al	13	Cu	29	0·153	0·304	49·4	5
Al	13	Pt	78	0·153	0·485	68·4	4
Cu	29	Zn	30	0·304	0·310	2·3	47
α-brass	29.4	β-brass	29·5	0·305	0·306	0·2	264

distances and large active detector areas, either scintillator or solid state, are desirable.

Although we may be able to detect two phases in a specimen, we may not be able to do so with very good spatial resolution, because, as we saw in section 5.5, and Figure 5.16, the ultimate resolution of the instrument is dependent on the contrast. Having calculated the atomic number contrast from equation 5.23, we can insert this value into equations 5.13 and 5.20, and hence calculate the spatial resolution of the two phases. This figure, shown in the last column of Table 5.1, should be considered as a degradation of the resolution, beyond that of $0 \cdot 1 \, \mu m$ due to the limitations of the sampling volume. It is very noticeable that for phases with similar atomic numbers the resolution may be very poor. Manufacturers of backscattered detectors will usually quote a figure for the atomic number difference that their equipment will detect, e.g. $0 \cdot 1$, but, without a figure for resolution under these conditions, this figure is not very meaningful.

With the advent of improved backscattered detectors over the past few years, use of this imaging mode in the SEM has increased greatly, particularly in conjunction with energy dispersive X-ray detection (Chapter 6). The areas of compositional interest are first identified by backscattered imaging and the phase compositions are then determined by spot analysis of the X-ray signal.

It is possible to obtain quantitative compositional information by measuring the intensity of the backscattered signal from the phase of interest, comparing it with a standard element, and then using Figure 5.8(a) to obtain the atomic number of the phase. Although this technique is potentially valuable on account of its high spatial resolution, and for its ability to analyse phases of low atomic number which are not easily determined using X-rays, great care must be taken to exclude other forms of contrast, such as topographic or crystallographic effects, if reliable results are to be obtained.

5.8 Crystallographic information from the SEM

We discussed in Chapters 3 and 4 how scattering of the primary electron beam by a specimen might be used to form diffraction patterns or to provide contrast in a TEM image. We can make use of similar electron beam interactions to obtain crystallographic information in the SEM.

5.8.1 Channelling contrast

The backscattered electron coefficient is dependent on the orientation of a crystal with respect to the incident beam. This effect, known as *electron channelling*, arises from the effects of diffraction on the depth of penetration of the primary beam into the specimen. The further the primary beam penetrates, the less likely are backscattered electrons to escape, and therefore the lower is the backscattered electron coefficient. Channelling contrast is generally

Figure 5.22 An example of channelling contrast in the backscattered image of a partly recrystallized aluminium alloy. Grains or subgrains of different crystallographic orientation
have different backscattering coefficients and are therefore imaged at different
grey levels.

much weaker than atomic number contrast, and may only be satisfactorily
obtained with a good backscattered electron detector, a carefully prepared
specimen which must not have a deformed surface layer such as is introduced
by mechanical polishing, and optimization of the microscope operating conditions. In particular, the electron beam must be reasonably parallel, and have
a large current.

Channelling contrast images are intermediate in spatial resolution between
optical and TEM images and are of particular use for the study of the grain
and subgrain microstructures developed during the processing of metals and
other crystalline materials. Because the contrast arises from Bragg diffraction,
only electrons which have lost little or no energy and which, therefore,
originate close to the point of impact of the electron beam (A in Figure
5.7(a)) contribute to the contrast and therefore the loss of spatial resolution
due to scattering within the sample (Figure 5.6) is quite low. The spatial resolution is optimum for low accelerating voltages and large atomic number. For
example, in aluminium at 8–10 keV the spatial resolution for a channelling
contrast image such as that of Figure 5.22 is ∼100 nm in a W-filament SEM
and ∼10 nm in a FEGSEM.

5.8.2 Diffraction patterns

Although we cannot obtain diffraction spot patterns in the SEM, as we can in the TEM, we can make use of electron channelling to form diffraction patterns. An early technique involved rocking the electron beam over a point on the specimen to produce 'selected area channelling patterns'. However, this technique has now been superseded by electron backscatter diffraction (EBSD) which is rapidly developing into an important analytical tool. The microscope arrangement for EBSD is shown schematically in Figure 5.23(a).

The specimen, which should be prepared as described above, so as to have a strain-free surface, is tilted to an angle of $\sim 70°$. The diffraction pattern, which is formed on a transmission phosphor screen, consists of lines, and, as shown in Figure 5.23(b), is similar to a Kikuchi pattern such as is obtained in the transmission electron microscope (Chapter 3 and Figure 3.18). The pattern is recorded using a sensitive video camera which is focused on the other side of the phosphor screen, and sent to a framestore in a PC. The raw pattern is noisy, has low contrast and is much more intense in the centre. Therefore before analysis, a background intensity is subtracted from the pattern and patterns from several video frames (typically 2–16) may be averaged to reduce the noise. The diffraction pattern is analysed in the PC, which is normally given information on the crystal structure and microscope operating conditions. The analysis program measures the positions of the lines and the angle between them and compares these to those predicted for the crystal structure. The crystallographic orientation of the sample is then calculated and the pattern stored. The time for acquisition and analysis of a diffraction pattern is $\sim 0·2\,s$ and the precision of angular measurement is typically $0·5–1·5°$. The spatial resolution for diffraction from aluminium in a W-filament SEM is $\sim 60\,nm$ in the direction parallel to the specimen tilt axis, although it is significantly worse in the orthogonal direction because of the asymmetry of the interaction volume (see Figure 5.17). However, using a FEGSEM, the resolution is improved in the same sample to $\sim 15\,nm$.

EBSD can therefore obtain diffraction data automatically and rapidly from very small volumes of material in bulk samples and can therefore perform some of the work which has traditionally been carried out by TEM. The orientations of a particular feature of the microstructure such as a recrystallized grain can be correlated with the microstructure, and by measuring the orientations of adjacent grains, grain boundary crystallography may be determined. Other applications include measurement of fracture plane, correlation of orientation and oxidation/corrosion behaviour and correlation of grain boundary crystallography with electrical, chemical or mechanical behaviour. As the sharpness or quality of the diffraction patterns is sensitive to the perfection of the microstructure, EBSD may be used to determine strain variations in a microstructure and the technique has been used to investigate the strains adjacent to crack tips.

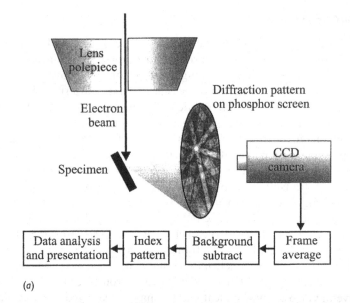

Lens polepiece

Electron beam

Specimen

Diffraction pattern on phosphor screen

CCD camera

| Data analysis and presentation | Index pattern | Background subtract | Frame average |

(a)

(b)

Figure 5.23 (a) Schematic diagram of the EBSD technique. The diffraction pattern from a highly tilted sample is detected on a transmission phosphor screen and recorded with a sensitive CCD camera. The pattern may be averaged over several TV frames and a background subtraction performed before the pattern is analysed by a PC. (b) An electron backscatter pattern from germanium. The pattern has a wider angular range (50°) than an SACP, and shows more detail. (D. J. Dingley)

Extensive data may be acquired by rastering the electron beam over the sample or alternatively moving the sample in a raster, whilst acquiring diffraction patterns. These data may be used to display 'maps' showing the orientation variation over a sample. These are analogous to the 'composition maps' which may be obtained using EDX or WDX (Chapter 6). Using beam scanning, up to 20 000 diffraction patterns, e.g. a grid of 200×100 points, may be collected in ~ 1 hr, or alternatively linear scans of the sample may be used.

Such data contain a very large amount of detailed information about the sample and can be used to determine grain and subgrain sizes and shapes, orientations and texture (Figure 5.24(d)) and misorientations (Figure 5.24(c)), phase distributions, etc. It is likely that EBSD will soon become a routine research tool for quantitative metallography.

Although EBSD is mainly used to determine the orientations of crystals as described above, it can also be used to determine the crystal structure of regions down to a few nanometres in diameter, and as this is achieved on bulk samples, there are some potential advantages over the TEM methods discussed in Chapter 3. The method is similar to that described above except that the diffraction pattern is recorded more slowly and at higher resolution, and the pattern line spacings and symmetry are compared with an extensive crystal database.

5.9 The use of other signals in the SEM

5.9.1 The charge collection mode

Every incident electron generates hundreds or even thousands of 'electron-hole pairs' when it knocks electrons from the outer shells of the atoms of the specimen, giving rise to a free electron and a 'hole' in the outer shell. Normally the vast majority of these pairs recombine within about 10^{-12} s – in other words the electrons jump back into their places in the shells extremely quickly. However, if the specimen is a semiconductor, and a voltage is applied across it, then the electrons and holes will be dragged apart before they can recombine, and a current will flow between the electrodes (Figure 5.25).

Alternatively, recombination of the carriers may be prevented by an internal field, such as that from a p–n junction. (It is this effect which is used in the solid state backscattered electron detector discussed in section 5.2 and the X-ray detector discussed in Chapter 6.) We can, using suitable equipment, measure either the current generated in the specimen, in which case we obtain an EBIC (electron beam induced current) signal, or we can measure the voltage induced by the beam – EBIV (electron beam induced voltage).

Using either of these signals we can display an image which will represent the variation of the semiconductor properties across the specimen as shown in Figure 5.26. The current flowing from each point will depend on the conduc-

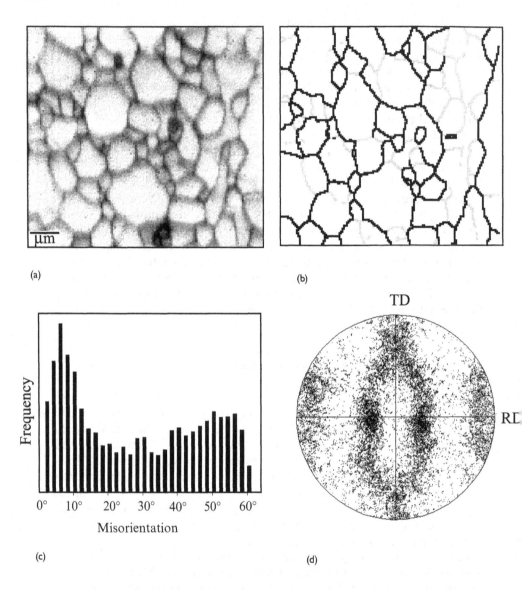

(a)

(b)

(c)

(d)

TD

RD

Frequency

0° 10° 20° 30° 40° 50° 60°

Misorientation

Figure 5.24 Analysis of an EBSD raster of 150 by 150 points with a step size of 60 nm on a fine-grained aluminium sample showing the extensive information available from such a data set. (*a*) A 'pattern quality' map constructed from the individual diffraction patterns. The intensity of each pixel is determined by the contrast in the diffraction pattern, and as this is low at the grain boundaries the resulting map is equivalent to an image of the microstructure. (*b*) The distribution of grain boundaries in the sample in which high angle grain boundaries are shown as black and low angle grain boundaries as grey. (*c*) The distribution of grain boundary misorientation angles in the sample. (*d*) A 111 pole figure constructed from the data.

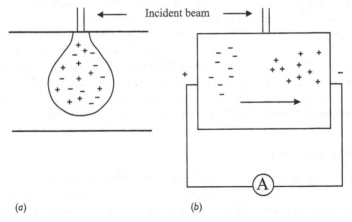

(a)   (b)

Figure 5.25 The 'charge collection' current mode of imaging in the SEM. (a) Electron hole pairs are generated within any specimen by the beam. (b) If a potential difference is applied across the specimen the electrons and holes can be separated before they recombine and a current will flow in the external circuit. This current is used as the signal for forming the image.

Figure 5.26 Scanning electron micrograph of a silicon IMPATT microwave frequency power generating diode. This is a double exposure micrograph in which secondary electrons were used to show the surface of the chip and the gold wire bonded to the aluminium metallized surface. The EBIC signal was used to show the Schottky barrier under the PtSi contact as a grey stripe surrounded by the p–n junction of the guard ring which appears white. (D. B. Holt, Imperial College, London)

tivity of the specimen at that point, the lifetime of electrons and holes, and their mobility (the drift speed under unit potential gradient).

These three parameters are of extreme importance to the semiconductor industry, and if their variation across a specimen can be measured or made visible in the SEM, then faults such as impurities, poor contacts etc. within a single integrated circuit can be investigated without damaging it (Richards and Footner, 1992). It is even possible to study an integrated circuit chip while it is being used, with currents flowing through its various components.

5.9.2 Cathodoluminescence

Many materials emit light under electron bombardment, and if this is detected, we can display an image in the *cathodoluminescent mode*. We have already encountered this effect in the phosphor used on the viewing screens of the transmission electron microscope and CRT tube of the SEM, and in the scintillator electron detectors used in the SEM. Cathodoluminescence (CL) varies in colour and intensity as a function of the composition of many minerals, and in semiconductors such as gallium arsenide (GaAs). Consequently this mode of imaging is of particular importance in these fields of application. Cathodoluminescence is also observed in polymeric or biological material, and in the latter materials, cathodoluminescent species may be used to 'label' the material.

Because cathodoluminescence may originate within most of the interaction volume of the sample, the spatial resolution is much worse than for electrons and is comparable with X-ray signals (Figure 5.6). However, the spatial resolution may be much improved by the use of low accelerating voltages and field emission electron sources.

As light detectors are usually also sensitive to electrons, CL detectors have to be designed to remove the electrons. Perhaps the simplest CL detection is achieved by removing the scintillating material from an electron detector so that the light is collected directly by the light guide. A specialized CL detector system is shown schematically in Figure 5.27. This employs a semi-ellipsoidal mirror which collects the light from the specimen and focuses it into a light guide and out of the microscope for spectral analysis.

CL appears as a band of wavelengths with a peak at a photon energy which is related to the energy gap between the filled valence band and empty conduction band of semiconducting and insulating materials. Any alteration in the energy gap due to local changes of temperature, crystal structure or impurity level will lead to a change in emission. At temperatures at or near liquid helium temperature, the CL emission bands generally become more intense and much sharper, and the band may be resolved into a line spectrum. Analysis of such a spectrum enables very low levels of impurity (< 0.01 ppm.) to be identified. Note that this is several orders of magnitude more sensitive than X-ray analysis (Chapter 6).

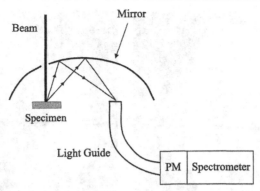

Figure 5.27 Schematic diagram of a cathodoluminescence detector which uses a semi-ellipsoidal mirror to collect the light and focus it into a light guide.

CL is being increasingly used to study the nature and distribution of defects in electronic materials and is capable of providing detailed and important information about the electronic effects of these defects. Figure 5.28 shows a high-resolution plan view CL image of an InGaN single quantum well structure grown on sapphire. Analysis of the light spectra from different features of such an image reveals differences in the electronic structure of

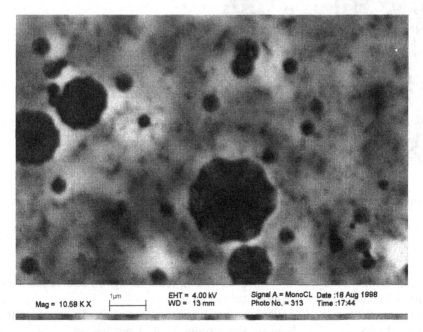

Figure 5.28 High resolution panchromatic CL image of an InGaN single quantum well structure grown on sapphire recorded at 4 keV and 100 pA current. (Courtesy of Oxford Instruments)

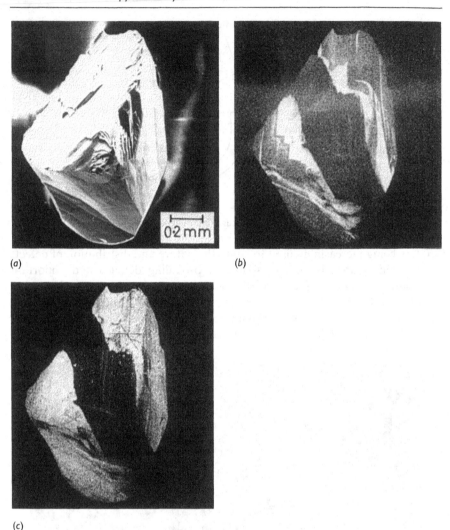

Figure 5.29 Micrographs of a natural diamond. (a) Secondary electron micrograph. (b) Monochromatic CL micrograph with 516 nm light. (c) Monochromatic CL micrograph with 444 nm light. (J. B. Steyn, P. Giles and D. B. Holt, Imperial College, London)

these regions. Interpretation of such CL data requires a good understanding of the solid-state physics of the material.

Figure 5.29 shows CL micrographs of a natural diamond. The light emitted from the specimen has been passed through a spectrometer, and the different wavelength emissions from different areas of the specimen can clearly be seen.

5.9.3 Use of other signals and contrast modes

Magnetic materials have their own contrast effects, because the magnetic field of a specimen will interact with the secondary and backscattered electrons. The SEM may therefore be used to determine the magnetic domain structure of the material. The *specimen current*, which is the current flowing to earth from the specimen, is sometimes used as a signal. As the specimen current is related to the number of electrons impinging on the specimen, less the number emitted as backscattered or secondary electrons, it gives a signal which is the inverse of the total electron emission. However, the advent of more efficient electron detectors has meant that this signal is now no longer of great interest for imaging, although it is usually monitored, because it enables the beam current to be measured.

The *transmitted electron beam* from a thin sample is, of course, the basis for transmission microscopy. Using a scintillator detector below the specimen in a SEM, or, alternatively, using a scanning system with a TEM, results in a *scanning transmission electron microscope or STEM*.

The resultant image from a STEM is similar to that from the same specimen in a TEM. The advantage of using STEM rather than TEM is that the image data are produced in serial form, and may therefore be readily processed (see section 5.10). Although this has found some application with low contrast materials, and those which degrade rapidly under the electron beam, this technique, when used in conjunction with a standard TEM or SEM is not widely used. However, purpose built or dedicated scanning transmission microscopes, with ultra high vacuum systems and field emission electron guns are powerful analytical instruments, and are discussed in Chapter 7.

5.10 Image acquisition, processing and storage

As mentioned earlier, many SEMs now use digital rather than analogue control and there are some major differences in the ways in which data are processed and recorded.

Having discussed, in the previous sections, how to set up the microscope, detector and specimen so as obtain the desired signal, we should pay some attention to how we will deal with the signal. One of the major differences between the TEM and the SEM is that in the latter we acquire the data gradually, pixel by pixel, and line by line, whereas in the former all the picture elements are built up together. Using computer terminology, we say that the SEM has *serial* data collection, and the TEM has *parallel* data collection.

The imaging data for the SEM comes as a varying electrical signal from the detector, and it is therefore easy to modify this signal with an amplifier.

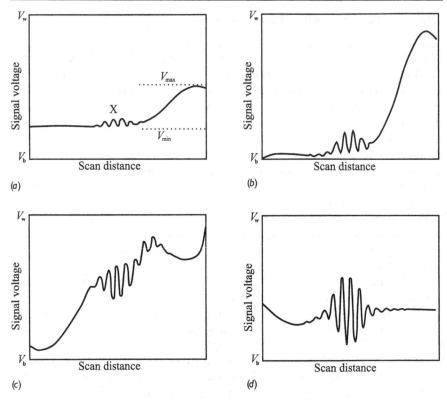

Figure 5.30 The simplest type of signal processing. (*a*) The original signal from the detector. (*b*) Signal adjusted for optimum display on CRT by increasing the amplifier gain (contrast control) and subtracting a DC or black level component (brightness control). (*c*) Non-linear amplification (gamma control) may be used to emphasize features with weak contrast. (*d*) Signal differentiation removes gradual background variations in the signal.

5.10.1 On-line processing

The simplest on-line processing involves the adjustment of image brightness and contrast using a linear amplifier. Controls for this will be found on all microscopes.

Consider the detector output voltage for a linescan which varies from V_{min} to V_{max} across a specimen, as shown in Figure 5.30(*a*), and note the voltages V_b and V_w required to give black and white on the CRT. In order for us to produce an image on the CRT which has intensity levels running from black to white, we need to adjust V_{min} until it equals V_b, and V_{max} until it equals V_w.

This is achieved by altering V_{max} (via the amplifier gain or contrast control) until $V_{max} - V_{min} = V_w - V_b$ and then subtracting a voltage V_{min} (the brightness control), until $V_{min} = V_b$, as shown in Figure 5.30(*b*).

In most circumstances this is all that is required to produce a good image, but some microscopes offer other signal processing options which may be of use in certain circumstances.

For example, if we wanted to emphasize the dark features 'X' in the image, we could use non-linear amplification. Instead of the amplifier output being proportional to the input, as it would be in a linear amplifier, we might arrange for V_{out} to be proportional to the square root of V_{in}. This would then amplify the dark features more than the light ones, as shown in Figure 5.30(c). This is called *gamma control*, after the term used to specify the grey level response of photographic film. Another type of processing sometimes used is *signal differentiation*, in which the output voltage of the amplifier is proportional to the slope of the voltage/time curve. This emphasizes regions where the signal changes rapidly, but flattens the contrast from a gradually changing background, as shown in Figure 5.30(d), and an example is shown for the channelling pattern in Figure 5.23(c).

It is also sometimes useful to add signals from different detectors or subtract them from each other as we saw for the backscattered images in Figures 5.19 and 5.21, and sometimes it may be beneficial to add or subtract two quite different signals.

It is easy to manipulate the incoming signals in a variety of ways, and to produce some very interesting results. However, the interpretation of the images may not be straightforward, and any but the simpler methods of signal processing should be used with caution.

5.10.2 Data handling with analogue microscopes

The microscope operator will need not only to view the scanned image so as to identify features of interest, focus and optimize the image, but also to obtain a permanent record of the chosen image. These two requirements often conflict.

We saw in section 5.5, that in order to produce a good quality, noise-free image from a given specimen, a minimum beam current is required, and that this is inversely proportional to the time for a frame scan F. Figure 5.16 shows how the minimum usable probe size, which is approximately equivalent to the spatial resolution for secondary electrons, varies with specimen contrast and frame scan time.

The best quality images are needed for photographic recording, which is carried out with a camera focused on a high resolution CRT. The scan rate is slowed right down so that one frame takes typically 50–100 s, while the camera shutter is open.

If the scan is carried out at a rate of 50 frames per second then the display appears simply as a television image. This is very convenient if the operator wants to move the specimen, focus the image, or simply to see what the specimen looks like. Also under these conditions, the microscope can be interfaced to a TV monitor or video recorder. However, as may be seen from Figure

5.16, the price that must be paid for such a fast scan rate is a dramatic loss of resolution. For example, considering a specimen with a contrast level of 50%, we see that although we might achieve a resolution of ~ 5 nm with a 100 s scan time recorded on photographic film, with a TV scan rate the resolution drops to ~ 100 nm. If we further assume that a TV system has a resolution of 500×500 pixels, then, using the arguments of section 5.4.1, it can be seen that the maximum useful magnification in these circumstances will be $\times 3000$.

If it is necessary to work at higher magnifications, so as for example to focus an image for photographic recording, then the operator must compromise, and work with a slower scan rate, perhaps one or two frames per second. This problem is even more acute when a detector with a slow response time, such as a solid state backscattered detector is used. In this case, fast scan rates cannot be used at any magnification.

5.10.3 Data handling with digital microscopes

In a modern digital SEM the beam is allowed to dwell on each point on the specimen (specimen pixel) for a pre-determined period and the image is built up by recording each pixel on an element of a *framestore*. The intensity of each pixel is typically stored as a number between 0 and 255, and the size of the image is controlled by the number of elements in the framestore, which can be varied, but is typically $\sim 1200 \times 1000$. (In the discussions of section 5.5, this figure replaces the resolution of the CRT which was taken to be 1000×1000 points.) The contents of the framestore are then displayed on a computer screen.

One advantage of such a system is that the framestore image is persistent, unlike that of a CRT, and therefore the operator can view the image without using TV scan rates which, as discussed above, give noisier images of lower resolution. With a framestore it is also possible to carry out *frame averaging*, in which the data from successive scans are averaged for each pixel. This is essentially the same as using a longer time per scan, and we can therefore replace the frame time F in equation 5.20 by $N \times F$, where N is the number of frames averaged. Figure 5.31 shows the effect of frame time on image quality.

A permanent record of the image in a digital SEM is obtained by saving the contents of the framestore to a computer as a bit-map image. Having obtained a digital image, a large variety of options are available for obtaining further quantitative data. These include measurement of phase volume fractions and the sizes and shapes of particles, grains, holes etc.

5.11 The preparation of specimens for examination in the SEM

Because in most cases the SEM is used to study the surface morphology, bulk specimens are normally used and specimen preparation is far simpler than for transmission electron microscopy. Since there are no lenses below the specimen

(a)

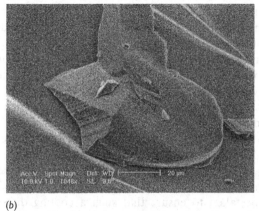

(b)

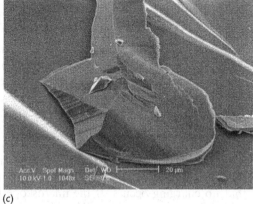

(c)

Figure 5.31 The effect of frame time on the quality of the secondary electron image of a fracture surface. (a) TV rate (20 ms), (b) 500 ms, (c) 25 000 ms.

there is a great deal of space available to accommodate the specimen and the various mechanical controls for manipulating it. This is particularly useful in the semiconductor industry, where it is necessary to inspect silicon wafers of diameter 200 mm or larger.

For effective viewing of a specimen in the SEM it is usually necessary for the surface of the specimen to be electrically conducting. This necessity arises from the statistics of electron yield. When a specimen is bombarded with high energy electrons, for each electron impinging on the specimen there is a yield of η backscattered electrons and δ secondary electrons. Figure 5.8(b) shows the variation of the total electron yield ($\eta + \delta$) with the accelerating voltage of the microscope. The implication of this curve is that there are only two operating voltages for the microscope where the yield is unity and hence electrons are leaving the surface at the same rate as they are hitting it. For most materials these two crossover voltages (E_1, and E_2) are in the range 1–5 keV, which is much lower than the normal operating voltage of the microscope. Consequently, during normal operation there is a surplus of electrons building up on the specimen surface. If these are not conducted away to earth, the specimen surface will become negatively charged until very soon the incoming primary electrons are repelled and deviated from their normal path, and a distorted image will be formed. Such charging effects are readily recognizable as seen in Figure 5.32(a).

Clearly there is no difficulty in studying clean metal specimens, provided that they are mounted so as to provide a conducting path to earth. However, non-conductors such as ceramics, polymers and biological materials present a problem. In this case it is usual to coat the specimen with a thin (~ 10 nm) conducting layer of gold or carbon, and this is nowadays easily and rapidly done by *sputter coating*. If the microscope is operating at the highest resolution, then care must be taken to ensure that such a coating does not mask fine surface detail. Also, coating may interfere with other signals or contrast modes, e.g. X-ray emission.

The examination of non-conducting specimens, particularly polymers and biological materials in the SEM may present other problems, such as specimen degradation by beam heating, radiation damage, and specimen volatility in the high vacuum. For further details of these effects, and the steps which may be taken to overcome them, the reader is referred to a more specialized text (see bibliography).

5.12 Low voltage microscopy

An alternative approach to the problem of specimen charging is to operate the microscope at a voltage where, according to Figure 5.8(b), the electron yield is close to unity, so that no charging occurs. If the microscope is operated at a voltage between E_1 and E_2 then the electron yield is greater than 1 and the sample charges positively. The potential between the filament and sample

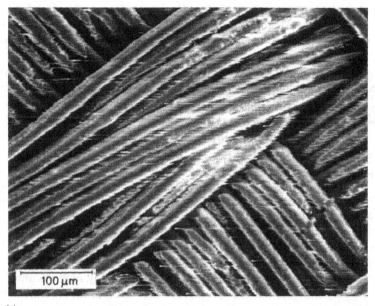

(a)

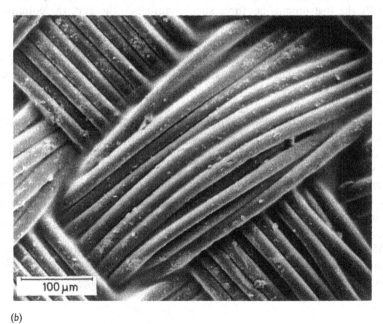

(b)

Figure 5.32 Secondary electron micrographs of a nylon fabric. (a) Uncoated specimen showing image degradation due to microdischarges. (b) A similar specimen after coating with 10 nm of gold in a sputter unit.

thus increases and the beam energy moves towards E_2. There is therefore a self-compensating effect which stabilizes the beam energy close to E_2. Because the scattering coefficients are strong functions of the surface tilt, it may not be easy to eliminate charging from all regions of the sample simultaneously.

The performance of conventional W-filament microscopes deteriorates markedly at low voltages, but modern FEG microscopes operate satisfactorily in the 1–5 keV range and may be used to examine uncoated ceramic and biological samples.

5.13 Environmental scanning electron microscopy (ESEM)

A normal SEM typically operates with a vacuum below 10^{-6} torr and samples associated with vapour or volatile components are unsuitable for examination without careful pre-treatment. Over the past few years, scanning electron microscopes which operate at pressures in the region of 1–10 torr have been developed to overcome these problems.

Such instruments operate by maintaining a reasonably high vacuum in the microscope column and a higher pressure in the region of the specimen. This is achieved by separating these regions by small apertures, often corresponding to the electron beam apertures, with each region having its own pumping system. Typically the instrument has three such chambers – the gun, the column and the specimen chamber.

Both secondary and backscattered electrons may be used in the ESEM. Large area scintillator detectors such as that shown in Figure 5.10(a) may be used to detect the backscattered electrons. The Everhart–Thornley detector cannot be used for secondary electron detection as the high bias voltage would cause electrical breakdown at high pressures. However, gas-phase detectors whose operation is analagous to the detectors used for wavelength dispersive X-ray detection (section 6.2.2) have been developed for secondary electron detection. Figure 5.33 is an example of the imaging of a wet sample in an ESEM using such a gas-phase detector.

There is generally some loss of resolution at high pressures due to elastic collisions between the electrons and the gas molecules. Nevertheless reasonable operation can be maintained. As well as enabling imaging of wet samples and allowing in situ experiments involving gas or liquid phases, ESEM allows the imaging of uncoated non-conducting samples because any accumulated electrical charges are dissipated by electron-molecule collisions.

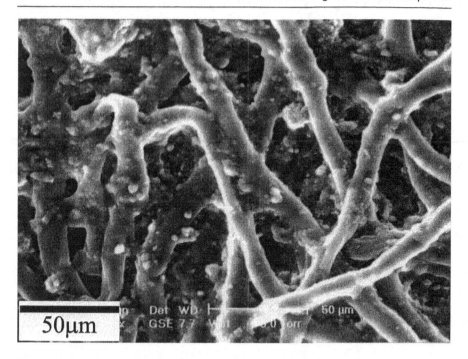

Figure 5.33 An algae mat in the fully wet condition imaged in an ESEM using a gas-phase secondary electron detector (courtesy of Philips).

5.14 Questions

1   Explain why reducing the probe size in the SEM results in a reduced beam current. Could this be countered by increasing the size of the objective aperture?

2   It is required to examine a fractured component in the SEM at a magnification of ×1000 and to achieve a depth of field of 1 mm. If the maximum working distance that can be used is 50 mm, what is the maximum sized objective aperture that can be used?

3   Explain why the electron probe size should be adjusted whenever the magnification is altered.

4   Calculate the natural contrast in a backscattered image between phases of mean atomic numbers 24 and 23.

5   If the natural contrast in a sample imaged using secondary electrons is 0·1, what is the best obtainable spatial resolution in a W-filament microscope operating with the following parameters using a frame time of 100 s?

Chapter 6

Chemical analysis in the electron microscope

It has been stressed in Chapter 2 that whenever electrons with several kilo electron volts of energy strike a solid specimen, X-rays characteristic of the atoms present in the specimen are produced. In discussing the formation of images in the TEM and SEM we have largely ignored these X-rays. However, to do so is to discard a great deal of information about the composition of the specimen. This was realized in the 1950s and since then increasing use has been made of all types of electron microscopes for *microanalysis*. This term implies that an analysis can be performed on a very small amount of material, or, more usually, a very small part of a larger specimen. As conventional chemical or spectrographic methods of analysis cannot do this, microanalysis in the electron microscope has become an important tool for characterizing all types of solid material.

In principle we can determine two things from the X-ray spectrum emitted by any specimen. Measurement of the wavelength (or energy) of each characteristic X-ray that is emitted enables us to find out which elements are present in the specimen, i.e. to carry out a *qualitative analysis*. Measurement of how many X-rays of any type are emitted per second should also tell us how much of the element is present, i.e. enable a *quantitative analysis* to be carried out. However, as this chapter will show, the instrumental and specimen requirements for quantitative analysis are such that the step from qualitative to quantitative analysis is not easily made.

We should note there are three types of electron microscopes commonly used for microanalysis. These are the SEM with X-ray detectors, the electron probe microanalyser, which is essentially a purpose-built analytical microscope of the SEM type, and transmission electron microscopes (TEM and STEM) fitted with X-ray detectors. The role of these different instruments will be discussed later in this chapter.

In a transmission electron microscope we can also obtain compositional information by measuring the energy loss of the transmitted electrons, and this technique will also be discussed.

6.1 The generation of X-rays within a specimen

We have seen in section 2.8 that bombardment of a material with high energy electrons will result in the emission of 'characteristic' X-rays, whose wavelengths depend on the nature of the atoms in the specimen, together with white radiation (or Bremsstrahlung) of all wavelengths down to a minimum corresponding to the incident electron energy (Figure 2.8). (The reader to whom these concepts are unfamiliar might find it helpful to re-read section 2.8 before continuing.) Before we can use the X-rays for analytical purposes we need to know which of the many characteristic X-ray lines for each element is the most intense; this enables us to choose the best line to use as an index of how much of each element is present in the sample. The situation at first appears to be very complex since, as Figure 6.1 shows, there is a large number of electron transitions possible in a large atom, each of which should lead to an X-ray of a unique wavelength.

It transpires, fortunately for the microanalyst, that in the K series the lines $K_{\alpha 1}$ and $K_{\alpha 2}$ (which may be so close together that they cannot be distinguished) are seven to eight times more intense than $K_{\beta 1}$ and $K_{\beta 2}$ (another close pair). Consequently the K 'doublet', as it is called, is most frequently used for analysis. However, it is not always possible to excite the K series of lines in an electron beam instrument since, as the atomic number of the emitting element increases, the energy required to knock out a K-shell electron also increases. For example, elements heavier than tin ($Z = 50$) need electrons of more than 25 keV to excite any K lines at all, and are not efficient producers of

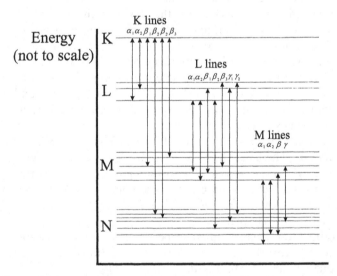

Figure 6.1 Some of the more common transitions between the K, L, M and N shells of an atom which lead to the X-ray lines indicated. The nomenclature is that of Figure 2.1.

K X-rays until the incident electron energy is about 75 keV. Since we would like to be able to analyse a specimen in an SEM, where electron energies of perhaps only 30 keV are available, we must look for other characteristic X-rays which are more easily excited in order to detect heavy elements. Fortunately the L series of lines shown in Figure 6.1, or even the M series for very heavy elements, has very suitable properties. Again it turns out that of the vast number of possible lines, $L_{\alpha 1}$ and $L_{\alpha 2}$ are far stronger than the remainder, which can normally be ignored. A similar effect narrows the M series down to a few useful lines. Table 6.1 shows the energy and associated wavelength of the strongest K, L and M lines of the elements. The most efficient production of X-rays generally occurs when the bombarding electrons have about three times the X-ray energy. A study of Table 6.1 will show that all elements have at least one strong X-ray line with energy less than 10 keV and therefore there should be no difficulty in analysing for all the elements, using a scanning electron microscope operating at 25–30 keV.

In addition to direct excitation by electrons there is a further mechanism of X-ray production which must be considered in microanalysis. X-rays which are passing through a specimen (perhaps having been generated previously by electron excitation) can themselves excite atoms which then emit characteristic X-rays of a slightly lower energy. Thus, for example, in a brass specimen the zinc K_α X-rays (energy 8·64 keV) can excite extra copper K_α X-rays whose energy is less (8·05 keV, see Table 6.1). This effect is known as *fluorescence*. It is not a very efficient process, since only a few per cent of the higher energy rays will successfully excite the lower energy radiation. However, it may significantly alter the relative amounts of characteristic radiation coming from alloys or compounds, particularly when elements with quite similar atomic numbers are present. In the example quoted we would expect that if the composition of the brass were 70% Cu/30% Zn we would find that rather more than the expected proportion of X-rays emitted were Cu K_α and rather fewer than expected were Zn K_α, because of the fluorescence effect. We shall see later that fluorescence is one of the effects which make accurate quantitative analysis difficult.

One of the key factors which determines the scale on which microanalysis can be carried out is the volume of the specimen which is penetrated by electrons (the interaction volume), and the volume of the specimen from which the detected X-rays originate (the sampling volume). As discussed in Chapter 5, and shown in Figure 5.6, because X-rays generated deep in the specimen can escape from the specimen and be detected, the sampling volume for X-rays is almost as large as the interaction volume. The interaction volume is smallest for low energy electrons and heavy elements, and it is difficult to make it much smaller than $1\,\mu m \times 1\,\mu m \times 1\,\mu m$ without reducing the energy of the electron beam (E_0) so much that no useful X-rays are emitted. As a consequence the smallest volume which it is practicable to analyse in a scanning microscope is $\sim 1\,(\mu m)^3$. This sounds like a small volume but it

Table 6.1 The energy and associated wavelength of the strongest K, L and M lines of the elements.

Element	Atomic Number Z	Relative Atomic Mass A_r	$K_{\alpha l}$ E (keV)	λ (nm)	$L_{\alpha l}$ E (keV)	λ (nm)	$M_{\alpha l}$ E (keV)	λ (nm)
			(a)	(b)	(a)	(b)	(a)	(b)
Hydrogen	1	1·0						
Helium	2	4·0						
Lithium	3	6·9	0·05					
Beryllium	4	9·0	0·11	11·40				
Boron	5	10·8	0·18	6·76				
Carbon	6	12·0	0·28	4·47				
Nitrogen	7	14·0	0·39	3·16				
Oxygen	8	16·0	0·52	2·36				
Fluorine	9	19·0	0·68	1·83				
Neon	10	20·2	0·85	1·46				
Sodium	11	23·0	1·04	1·19				
Magnesium	12	24.3	1·25	0·99				
Aluminium	13	27.0	1·49	0·83				
Silicon	14	28·1	1·74	0·71				
Phosphorus	15	31·0	2·01	0·61				
Sulphur	16	32·1	2·31	0·54				
Chlorine	17	35·5	2·62	0·47				
Argon	18	39·9	2·96	0·42				
Potassium	19	39·1	3·31	0·37				
Calcium	20	40·1	3·69	0·34	0·34	3·63		
Scandium	21	45·0	4·09	0·30	0·39	3·13		
Titanium	22	47·9	4·51	0·27	0·45	2·74		
Vanadium	23	50·9	4·95	0·25	0·51	2·42		
Chromium	24	52·0	5·41	0·23	0·57	2·16		
Manganese	25	54·9	5·90	0·21	0·64	1·94		
Iron	26	55·8	6·40	0·19	0·70	1·76		
Cobalt	27	58·9	6·93	0·18	0·77	1·60		
Nickel	28	58·7	7·48	0·17	0·85	1·46		
Copper	29	63·5	8·05	0·15	0·93	1·33		
Zinc	30	65·4	8·64	0·14	1·01	1·23		
Gallium	31	69·7	9·25	0·13	1·10	1·13		
Germanium	32	72·6	9·88	0·12	1·19	1·04		
Arsenic	33	74·9	10·54	0·12	1·28	0·97		
Selenium	34	79·0	11·22	0·11	1·38	0·90		
Bromine	35	79·9	11·92	0·10	1·48	0·84		
Krypton	36	83·8	12·65	0·10	1·59	0·78		
Rubidium	37	85·5	13·39	0·09	1·69	0·73		
Strontium	38	87·6	14·16	0·09	1·81	0·69		
Yttrium	39	88·9	14·96	0·08	1·92	0·64		
Zirconium	40	91·2	15·77	0·08	2·04	0·61		
Niobium	41	92·9	16·61	0·07	2·17	0·57		
Molybdenum	42	95·9	17·48	0·07	2·29	0·54		
Technetium	43	98·0	18·36	0·07	2·42	0·51		
Ruthenium	44	101·1	19·28	0·06	2·55	0·48		
Rhodium	45	102·9	20·21	0·06	2·70	0·46		

Table 6.1 (continued)

Element	Atomic Number Z	Relative Atomic Mass A_r	$K_{\alpha I}$ E (keV)	λ (nm)	$L_{\alpha I}$ E (keV)	λ (nm)	$M_{\alpha I}$ E (keV)	λ (nm)
			(a)	(b)	(a)	(b)	(a)	(b)
Palladium	46	106·4	21·17	0·06	2·70	0·44		
Silver	47	107·9	22·16	0·06	2·98	0·41		
Cadmium	48	112·4	23·17	0·05	3·13	0·39		
Indium	49	114·8	24·21	0·05	3·29	0·38		
Tin	50	118·7	25·27	0·05	3·44	0·36		
Antimony	51	121·7	26·36	0·05	3·60	0·34		
Tellurium	52	127·6	27·47	0·04	3·77	0·33		
Iodine	53	126·9	28·61	0·04	3·94	0·31		
Xenon	54	131·3	29·77	0·04	4·11	0·30		
Caesium	55	132·9	30·97	0·04	4·29	0·29		
Barium	56	137·3	32·19	0·04	4·46	0·28		
Lanthanum	57	138·9	33·44	0·04	4·65	0·27	0·83	1·49
Hafnium	72	178·5	55·78	0·02	7·90	0·16	1·64	0·75
Tantalum	73	181·0	57·52	0·02	8·14	0·15	1·71	0·73
Tungsten	74	183·8	59·31	0·02	8·40	0·15	1·77	0·70
Rhenium	75	186·2	61·13	0·02	8·65	0·14	1·84	0·67
Osmium	76	190·2	62·99	0·02	8·91	0·14	1·91	0·65
Iridium	77	192·2	64·88	0·02	9·17	0·14	1·98	0·63
Platinum	78	195·1	66·82	0·02	9·44	0·13	2·05	0·60
Gold	79	197·0	68·79	0·02	9·71	0·13	2·12	0·58
Mercury	80	200·6	70·81	0·02	9·99	0·12	2·19	0·56
Thallium	81	204·4	72·86	0·02	10·27	0·12	2·27	0·55
Lead	82	207·2	74·96	0·02	10·55	0·12	2·34	0·53
Bismuth	83	209·0	77·10	0·02	10·84	0·11	2·42	0·51
Polonium	84	210·0	79·28	0·02	11·13	0·11	?	?
Astatine	85	210·0	81·50	0·02	11·43	0·11		
Radon	86	222·0	83·77	0·01	11·73	0·11		
Francium	87	223·0	86·09	0·01	12·03	0·10		
Radium	88	226·0	88·45	0·01	12·34	0·10		
Actinium	89	227·0	90·87	0·01	12·65	0·10		
Thorium	90	232·0	93·33	0·01	12·97	0·10	3·00	0·41
Proactinium	91	231·0	95·85	0·01	13·29	0·09	3·08	0·40
Uranium	92	238·0	98·42	0·01	13·61	0·09	3·17	0·30

contains about 10^{11} atoms! The sampling volume will also depend on the energy of the X-ray being emitted, because an X-ray of energy E_c will only be emitted if the electron energy E is greater than E_c. Thus, the maximum depth from which X-rays originate is effectively the depth at which the electron energy falls below E_c. Several attempts have been made to calculate the depth of X-ray production (R), and one commonly used approximation is that R is given (in micrometres, when E_o and E_c are in keV) by

$$R = P(E_o^{1.68} - E_c^{1.68}) \qquad (6.1)$$

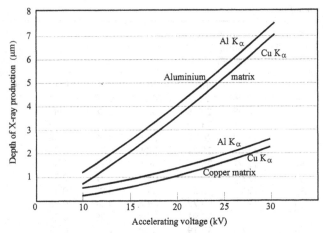

Figure 6.2 The depth from which X-rays are produced in a specimen as a function of the electron energy, the nature of the X-rays being detected, and the composition of the matrix.

where P is about 10^{-2} for materials with medium average atomic mass. To illustrate this point, Figure 6.2 shows the depth from which $Cu\,K_\alpha$ and $Al\,K_\alpha$ X-rays may be produced by electrons in an Al–Cu alloy in which the major element is either aluminium or copper.

If X-ray fluorescence occurs, then the sampling volume may actually be larger than the interaction volume.

The fraction of the generated X-rays which reach the specimen surface and are emitted will depend both on the energy of the X-rays and the average atomic weight of the specimen. For example, soft (low energy, long wavelength) X-rays such as carbon K_α are readily absorbed by solids and therefore relatively few escape from the surface. On the other hand, hard (high energy, short wavelength) X-rays such as molybdenum K_α can penetrate many micrometres of most solids, and are depleted only a little by absorption in the specimen. Clearly therefore both the volume which is being analysed, and the fraction of the X-rays which are emitted from the specimen depend very critically on (*a*) the energy of the electron beam, (*b*) the energy (wavelength) of the X-ray being studied and (*c*) the local atomic weight of the specimen.

As was stressed earlier, this complexity makes accurate analysis extremely difficult.

6.2 Detection and counting of X-rays

Ideally a system which is to be used for the analysis of a specimen in a scanning or transmission electron microscope should be able to reproduce for us the entire spectrum of X-rays emitted from the surface. However, as we have by now learnt to expect, the 'ideal' is rarely attainable in practice. In the present case we generally have to choose between two very different methods of obtaining the data. Parts of the spectra obtained using these techniques, from the silver solder joint whose microstructure is shown in Figure 5.21, are shown in Figure 6.3. The *wavelength dispersive* spectrometer or WDS is the basis of the purpose-built electron probe microanalyser. The spectrum obtained with a wavelength dispersive spectrometer is shown in Figure 6.3(a). These devices can determine extremely accurately the position of a single X-ray line (i.e. its wavelength or energy), can resolve closely spaced lines and are particularly suited to measuring the height (intensity) of a peak above the background level. Alternatively, many SEMs and TEMs are equipped with an *energy dispersive spectrometer* or EDS detection system which is able to detect and display most of the X-ray spectrum, but with some loss of precision and resolution as indicated in Figure 6.3(b). Nowadays, the distinction between analytical scanning microscopes and electron probe microanalysers is not clear, as some SEMs are fitted with WDS systems and some microanalysers use EDS as well as WDS. We shall now consider the two types of analysis system in more detail, to see why there is still a need for both.

6.2.1 Energy-dispersive analysis

We shall deal with the energy-dispersive detection system first, since, although it is historically the more recent, it is now the more generally applicable and certainly the more versatile system. The development of EDS analysis has been responsible for a major revolution over the past twenty five years in the use of electron beam instruments for the microcharacterization of materials.

In outline, the detector normally consists of a small piece of semiconducting silicon or germanium which is held in such a position that as many as possible of the X-rays emitted from the specimen fall upon it. Since X-rays cannot be deflected, the detector must be in the line of sight of the specimen. This means that in a scanning electron microscope it normally occupies a similar position to the secondary electron detector (see Figure 5.9). In order to collect as many X-rays as possible the silicon should be as near to the specimen as is practicable. In a SEM it may be possible to place the detector 20 mm or less from the specimen, but the problems are greater with a TEM because the specimen is within the objective lens.

The detector works in the following way. Each incoming X-ray excites a number of electrons into the conduction band of the silicon leaving an identical number of positively charged holes in the outer electron shells. The energy required for each of these excitations is only 3·8 eV; consequently the number of electron-hole pairs generated is proportional to the energy of the X-ray photon being detected. For example, an Al K_α X-ray, with an energy of

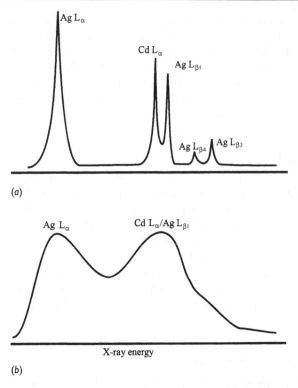

Figure 6.3 Part of the spectra obtained from a specimen of silver solder, a copper–silver–cadmium alloy. (*a*) Spectrum obtained using wavelength dispersive analysis. (*b*) The same region of the spectrum, obtained using energy dispersive X-ray analysis. Using this technique, the Ag L_β and Cd L_α peaks are not resolved and the Ag L_α and Cd L_α are barely resolvable.

1·49 keV, will give rise to approximately 390 electron-hole pairs. If a voltage is applied across the semiconductor a current will flow as each X-ray is absorbed in the detector and the magnitude of the current will be exactly proportional to the energy of the X-ray. In practice, if pure silicon is used the current generated is minute compared with the current which flows normally when a voltage is applied; in other words the resistivity is too low. This is overcome by three stratagems which combine to make the final detector seem rather more complicated than it really is. Figure 6.4 shows the result, and contains the features which are common to virtually all energy-dispersive detectors.

The resistivity of the silicon is increased by (a) making the whole detector a semiconductor p–i–n junction which is reverse biased, (b) doping the silicon with a small concentration of lithium, and (c) cooling the whole detector to liquid nitrogen temperature (77 K). The current which normally passes between the gold electrodes is now very small indeed until an X-ray enters the detector, then the resultant current can be amplified and measured fairly easily. The

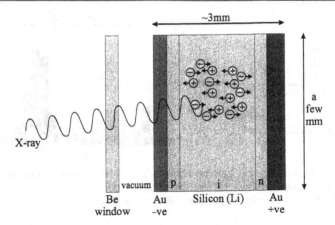

Figure 6.4 A silicon energy-dispersive X-ray detector. The beryllium window and gold contact layers are grossly exaggerated in thickness; typical thicknesses would be 7–8 μm for the Be, and 10–20 nm for the Au.

detector shown schematically in Figure 6.4 consists of a Si(Li) semiconductor junction in which the i region occupies most of the 3 mm thickness. Thin layers of gold are necessary on both surfaces of the detector so that the bias potential can be applied. The film of gold on the outer face of the detector must be as thin as possible so that very few X-rays are absorbed in it; a layer only 20 nm thick provides adequate conductivity. The gold-coated outer surface is usually further protected by a thin window of beryllium or a polymer. This window is necessary to prevent contaminants from the specimen chamber of the micro-scope from condensing on the very cold surface of the detector and forming a further barrier to the entry of X-rays. Unfortunately the window itself, despite being made of beryllium ($Z = 4$) or carbon ($Z = 6$) and only being a few micrometres thick, absorbs a significant proportion of the low energy X-rays falling on the detector and therefore makes light elements particularly difficult to detect. It is impracticable to detect X-rays of energy less than 1 keV with this type of detector, and this therefore eliminates all elements lighter than sodium. Windowless detectors, or detectors with ultra-thin windows of formvar or some other polymer film are now available, and these extend the analytical range down to boron. However, such detectors need to be used in microscopes which have extremely good vacuum systems, and need very careful protection from accidental air leaks such as those that occur during specimen changes.

The current which flows between the electrodes when an X-ray enters the detector lasts for an extremely short time (less than 1 μs) and is normally referred to as a *pulse*. Each pulse is amplified and then passed to a computer acting as a multichannel analyser (MCA), which decides which of perhaps 1000 channels, each representing a different X-ray energy, the pulse should be regis-tered in. The MCA thus effectively collects a histogram of the energies of all the

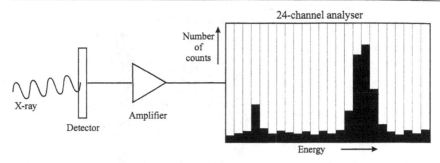

Figure 6.5 A simplified energy-dispersive analysis system. Pulses from the detector are amplified and then stored in the appropriate channel of multi-channel analyser. In reality, the MCA would contain perhaps 1000 channels instead of the 24 shown here.

X-rays arriving at the detector, as indicated in Figure 6.5. This histogram is then displayed on a screen, and usually appears as a smooth curve such as is shown in Figure 6.3(*b*).

The solid state detector is very efficient, and almost 100% of the X-rays entering the detector crystal will produce a pulse. However, the pulse processing time, during which an X-ray photon is detected and the resulting pulse amplified and sorted by the MCA, is short but finite, and this process must be completed before the next pulse can be dealt with. The pulse pile-up rejection circuit manages this operation, which is a very critical aspect of the EDS system. Pulse processing speed limits the rate at which X-rays can be counted, and at the time of writing this is typically about ten thousand counts per second (cps). If the count rate is less than a few thousand cps, then most of the incoming pulses are processed, but as the count rate rises, an increasing fraction of the pulses are rejected. The total time elapsed during an analysis consists of the time for which the detector was counting (*live time*), plus the time for which the detector was ignoring incoming X-rays (*dead time*). In determining the X-ray count rate from a specimen, it is therefore the number of counts collected for a given *live time*, not *elapsed time*, which is required. Fortunately, modern EDS systems will record both times, and this should cause the analyst no problems.

The high efficiency of the EDS detector, coupled with the relatively large collection angle (typically greater than 0·5 steradians in an SEM) means that data may be collected rapidly at quite low beam currents. From a typical sample, a reasonable spectrum can usually be obtained within a minute or two.

The EDS system is controlled by a computer which also stores the energies of the X-rays from all the elements; consequently it is a simple matter for the computer to identify the element giving rise to a line in the spectrum, or alternatively to indicate on the screen the positions on the spectrum at which lines for any chosen element would appear. Thus qualitative analysis is extremely rapid with such a system. Figure 6.6 shows typical spectra

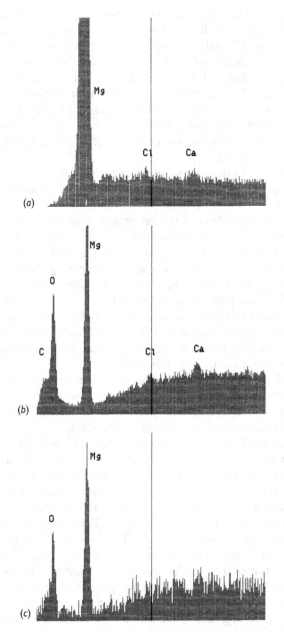

Figure 6.6 EDS spectra obtained from a particle of impure MgO on a carbon support film. The vertical line in the centre of the spectrum is the computer cursor from which the operator can read off the energy of any channel of the spectrum. (a) Spectrum from a standard detector with a beryllium window. (b) Spectrum obtained in 300 s using an ultra-thin window, light element detector, (c) as (b) but with a 50 s count time.

obtained from impure MgO particles on a carbon substrate. The microscope was equipped with a normal Be window detector as well as a thin-window detector for light element analysis, and the difference in performance of the detectors at the light element end of the spectrum can be clearly seen.

From the foregoing description it would seem that the EDS is an ideal system for presenting all the available X-ray data from a specimen in a convenient form from which both qualitative and quantitative analyses could be deduced. However there are some serious limitations to be considered.

First as discussed above, detection of elements lighter than sodium is impossible with a standard detector.

Secondly, in order to both preserve the detector crystal and to reduce noise in the system, the detector must be kept at 77 K at all times, which is clearly an experimental difficulty.

Thirdly, the energy resolution of the detector is poor; in other words, each X-ray line is not detected as a sharp line, but, as seen in Figure 6.6, as a broad peak, typically 100–200 eV wide. Not only does this make it impossible to resolve closely spaced lines, as may be seen by comparing the Ag_L and Cd_L lines in Figures 6.3(a) and 6.3(b), but, as any X-ray line now occupies several channels of the MCA, the peak height is reduced. This factor, together with the relatively large amount of electronic noise in the system results in rather low peak to background ratios compared to WDS. As we shall see later, a low peak to background ratio will affect not only the limit of detectability of the analyser, but also its use for quantitative analysis.

A further problem with an energy dispersive system is that under certain circumstances spurious peaks in the spectrum may be produced. The most important of these effects are the *sum peak* and the *silicon escape peak*. As discussed above, the energy of an X-ray photon entering the detector is determined from the number of electron-hole pairs created. If two identical photons enter the detector simultaneously, then twice as many hole pairs are created, the detector system will interpret this as being due to a single photon of twice the energy, and this pulse will be sorted accordingly to produce a sum peak at an energy of twice that of the X-rays being detected. The escape peak is due to an X-ray entering the detector and ionizing a silicon atom in the detector crystal by knocking out a K-shell electron. The original X-ray photon loses 1·74 keV in energy (i.e. the energy of SiK_α X-rays), and unless the resulting SiK_α X-ray photon is absorbed by the detector, thus releasing its energy to the detector, then a peak 1·74 keV below the main peak is produced.

Both escape peaks and sum peaks are only significant under conditions where there is a very strong main peak and a high count rate. As they occur in predictable positions in the spectrum, they are easily recognized by an experienced microscopist, and by the computer.

Having painted such a gloomy picture of the energy-dispersive analysis system we should emphasize the advantages. With an energy-dispersive system we can place a detector very close to the specimen in a TEM or SEM and therefore collect the X-rays very efficiently, and as X-rays of all energies are collected at the same time we can acquire a complete spectrum, and therefore a qualitative analysis, in a few minutes.

6.2.2 Wavelength-dispersive analysis

The three areas in which the EDS system performs badly – light element detection, peak separation, and peak to background ratio – are the strong points of the other major X-ray detection system for electron microscopes, the wavelength-dispersive spectrometer (WDS).

The principle of the WDS is that the X-radiation coming from the specimen is filtered so that only X-rays of a chosen wavelength (usually the characteristic wavelength of the element of interest) are allowed to fall on a detector. The 'filtering' is achieved by a crystal spectrometer which employs diffraction to separate the X-rays according to their wavelength. A common arrangement for the spectrometer is illustrated in Figure 6.7. The X-rays leaving the specimen at a certain angle, the *take-off angle* (ϕ), are allowed to fall onto a crystal of lattice spacing d. If the angle between the incident X-rays and the crystal lattice planes is θ, then the only X-rays which will be diffracted by the crystal and thus reach the detector will be those obeying Bragg's Law (section 3.1). The wavelength of the transmitted X-rays is therefore given by

$$\lambda = \frac{2d \sin \theta}{n} \tag{6.2}$$

If the spectrometer is to be used to detect, say $Fe\,K_\alpha$, then the characteristic wavelength (0·19 nm from Table 6.1) is substituted into equation 6.2, and the appropriate value of θ can be calculated for the particular crystal in use. If the spectrometer is set to this angle then only the $Fe\,K_\alpha$ characteristic X-rays will reach the detector and be counted. The detector no longer has to discriminate between X-rays of different energies – it is sufficient for it to count the X-ray photons arriving. A much simpler detector than is found for EDS is used. This is the gas proportional counter, in which much faster count rates can be tolerated. One of the disadvantages is that because the X-rays must be reasonably well collimated before reaching the crystal, two sets of slits (S_1 and S_2) are generally used. In order that as many as possible of the X-rays leaving the crystal arrive at the detector the geometry of the spectrometer is chosen so that, as shown in Figure 6.7, all possible X-ray paths are focused onto the detector.

In order to achieve this focusing effect the specimen, crystal and detector must all lie on a circle of radius R known as the Rowland circle and shown dotted in Figure 6.7. Also the crystal must be curved – in the Johan type of spectrometer shown in the figure it is bent to a radius $2R$. These requirements place several restrictions on the analysis system: the spectrometer is necessarily quite large; its mechanism is complicated, since it has to be able to alter θ while keeping both the crystal and the detector on the Rowland circle; precision

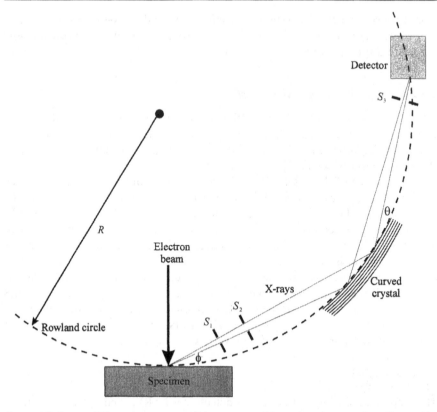

Figure 6.7 A crystal X-ray spectrometer. X-rays emitted from the specimen are collimated by two slits S_1 and S_2, diffracted by the curved crystal, and then focused on to the detector. For the maximum efficiency the specimen, crystal and detector must all lie on the Rowland circle of radius R.

engineering is required, since, in order to discriminate between very close lines, the angle θ must be set to an accuracy of better than one minute of arc; finally the position of the specimen is absolutely critical, since if it lies off the Rowland circle by as little as a few micrometres the number of X-rays reaching the spectrometer will be severely reduced. These restrictions make the design of a crystal spectrometer difficult and expensive. The mechanics of a crystal spectrometer allow the angle θ to be altered only over a limited range, and therefore to cover the whole X-ray spectrum a range of crystals, normally four, with different lattice spacings, are required. Although some spectrometers are designed to hold only one crystal, requiring the instrument to be equipped with four spectrometers, many spectrometers now have two or four crystals which may be interchanged automatically.

Because of the design of a crystal spectrometer, the geometric efficiency is low, and a solid collection angle of 0·01 steradians is typical. The detector efficiency is also quite low, typically less than 30%, because of losses both in the diffracting crystal and at the counter. Despite the difficulty and expense of a crystal spectrometer, it remains an essential tool for analysis in the electron microscope because of four major advantages:

(a) The resolving power of a crystal spectrometer is excellent. As may be seen in Figure 6.3(a), the lines are sharp and there is rarely a problem of adjacent lines overlapping.

(b) The peak-to-background ratio of each line is much higher (often by a factor of ten) than can be achieved with a solid detector.

(c) Since the X-ray detector normally used in a crystal spectrometer is capable of counting at very high rates (perhaps 50 000 cps) it may be possible to collect data from a single element in a very short counting time.

(d) One of the main reasons for using a crystal spectrometer is its ability to detect X-rays from light elements. With suitable crystals of large lattice spacing it is possible to detect and count X-rays as soft as boron $K\alpha$ or even beryllium $K\alpha$.

Although, as mentioned above, a windowless EDS detector can be used down to the limit of boron, the greater sensitivity and better resolving power of the crystal spectrometer give this technique a distinct advantage at the present time.

On the other hand, the disadvantages of an X-ray spectrometer make its use very time-consuming and hence restrict its major applications to the things it does well – detection of low concentrations, detection of light elements, and quantitative measurement of peak heights. There are some obvious disadvantages in using a crystal spectrometer for qualitative analysis. These are that as only one wavelength of X-ray is detected at any time, a scan of all the required wavelengths can take a considerable time. Also, since the spectrometer employs diffraction there will be not one angle of the crystal (θ) at which a certain line is detected but several, one corresponding to each value of n in equation 6.1. Thus there may be as many as seven or eight *orders of reflection* detectable from a major X-ray line and therefore the spectrum contains far more lines than that collected by the energy-dispersive detector.

In summary, the two main X-ray detection and measurement systems, EDS and WDS, each have their strong points. The EDS is clearly better suited to rapid qualitative analysis, while the WDS may give more accurate quantitative results. As we shall see in the next section, both have wide application in electron microscopy.

6.3 X-ray analysis of bulk specimens

6.3.1 Instrumentation

The earliest electron probe microanalysers pre-dated the SEM and were simply instruments which generated a beam of electrons, directed it at a specimen and used one or more crystal spectrometers to analyse the X-radiation emitted. This represented a great advance in analytical technique, since the electron beam could be directed at a relatively small area of a solid specimen and an analysis obtained without destroying the specimen. The region of the specimen to be analysed was selected using a light microscope, and the accuracy with which the beam could be positioned was rather limited. It was soon realized that by scanning the electron beam and installing an electron detector, an imaging system could be incorporated. Since the size of the sampling volume discussed in section 5.1 implies that no useful analysis can be made of a region smaller than $1\,\mu m \times 1\,\mu m \times 1\,\mu m$ there seemed to be no point in making the electron beam smaller than this. Indeed, as the low geometric efficiency of a crystal spectrometer means that a probe current of the order of $1\,\mu A$ is required, the arguments of section 5.5, and equation 5.13, show that a probe of diameter $\sim 1\,\mu m$ is needed to obtain this level of beam current.

However, electron probe microanalysers have progressed considerably, and an example is shown in Figure 6.8. A typical instrument is equipped with four computer controlled crystal spectrometers, containing a range of crystals such that the whole spectrum can be covered. An EDS system for preliminary qualitative analysis is fitted. Secondary and backscattered imaging can be carried out, and the spatial resolution in secondary electron mode is better than 10 nm. A fixed focus light microscope is retained, not only for examination of the specimen, but more importantly for ensuring that the specimen height may be adjusted until it is on the Rowland circle. In the design of such an instrument, particular attention is paid to the electron gun which must be capable of providing the large beam currents necessary coupled with the extreme stability which is required if an analysis is to be carried out over a period of several hours.

As well as automation of the spectrometers, the specimen stage will be motorized and be capable of computer control. Thus, areas of interest may be selected by the operator, and the instrument may then be left to carry out the analysis, perhaps running for several hours under automation. Such an instrument is ideally suited to an environment where accurate quantitative analysis is required from a large number of specimens on a routine basis.

An alternative approach is to mount an energy dispersive detector on to an existing scanning microscope. This is ideal for qualitative analysis, and the majority of scanning electron microscopes are now equipped with an EDS system, an example being seen in Figure 5.3. For reasons which are discussed in section 6.5, this may not be an ideal combination for the purposes of quantitative analysis; nevertheless, some scanning microscopes

Figure 6.8 An electron probe microanalyser. (JEOL Ltd.)

are capable of quantitative analysis, and some may also be fitted with crystal spectrometers. For a laboratory which does not have the requirement for a dedicated electron probe microanalyser, a good analytical SEM provides an acceptable and cheaper alternative.

6.3.2 The presentation of analytical information

For all the instruments discussed in this section, which are based on a scanning electron beam, there are various ways in which qualitative analytical information obtained from a suitable (ideally, flat) specimen can be presented, and the method selected will depend upon which of three types of information is required. This will be either (*a*) an analysis of a selected region of chemically homogeneous composition such as a phase, (*b*) an overall analysis of the whole or part of a sample, or (*c*) an analysis to determine the variation of composition within a region of the sample.

The first type of analysis is usually accomplished simply by a 'spot analysis', in which the electron beam is stopped and positioned carefully on the point to be analysed, which has been selected on the SEM screen while the image was still being displayed. The X-ray data, either a single peak or series of peaks being detected by a crystal spectrometer, or the whole spectrum being accumulated by an EDS system, can be collected for as long as is necessary and the composition at the selected point (or more correctly in the sampling volume of about $1\,\mu m^3$) can be determined. As an alternative, instead of using a stationary beam, we might, while analysing, scan a small raster on the specimen, making sure that the raster was contained within a chemically homogeneous region.

The second type of analysis entails obtaining data from a larger area of the specimen, the actual area being, of course, dependent on the scale of the chemical heterogeneity in the specimen. For an EDS system this only requires scanning the beam over a suitable area which might be $100\,\mu m \times 100\,\mu m$, as the X-rays are detected. The analysis is then an average of the area of the image on the screen. However, for a WDS system this is not possible, because, as may be seen from Figure 6.7, only one point P (strictly, one line) of the specimen lies on the Rowland circle. Therefore as the beam scans a raster, X-rays from regions other than those within a few micrometres of P will not be focused on the detector, will not be collected effectively, and a misleading result may be obtained. If an analysis of an area greater than about $5\,\mu m \times 5\,\mu m$ is required by WDS, then this must be done by scanning the specimen and not the beam.

In order to examine the variation of chemical composition within a sample, many analyses must be carried out over the area of interest. There are three approaches to this problem; the simplest, and probably the best, is to use backscattered electron compositional imaging (section 5.7) to identify regions of chemical homogeneity, and then to carry out spot analyses of these areas.

An alternative is to select the X-ray signal from an element of interest, and to display its intensity as the beam (or for a WDS system, the specimen) is scanned. For a crystal spectrometer, this simply requires the use of a *ratemeter* which will measure the instantaneous count rate, rather than the total number of counts from the selected peak. For an EDS system, it is necessary to make the ratemeter sensitive only to the counts going to a selected number of channels. If the X-ray count rate measured by the ratemeter is used to deflect a spot on the display screen then a trace of composition versus distance as illustrated in Figure 6.9 can be obtained. Notice that although for ease of recognition the 'line' of the analysis is shown superimposed on a micrograph of the specimen, these were taken separately. First an electron image of the specimen was recorded, then the spot was scanned along the chosen line and recorded on the same photographic exposure. Then the spot was scanned very slowly, taking 100–500 s, while the trace shown in Figure 6.9 was displayed and photographed.

As can be seen from the example illustrated, this technique makes it quite simple to find the regions where there is a significant change in the concentration of an element. An accurate analysis is difficult, as the electron beam spends only a short time on each spot, and the counting statistics are poor.

Another way of displaying information about the distribution of a single element is by what is often termed X-ray mapping. Using this technique, the X-ray counter is used in a similar way to any of the other detectors in the SEM to

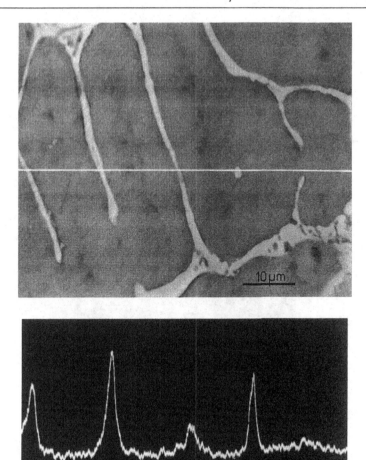

Figure 6.9 The variation of copper concentration across a section of an aluminium–copper alloy. The trace in the lower part of the figure is the copper count rate (i.e. the output of the ratemeter) as the electron beam was scanned along the line indicated in the micrograph.

form an image. This is essentially an extension of the linescan method discussed above, to two-dimensional scanning. In its simplest form, the display is made bright every time an X-ray photon is counted. The image then consists of bright dots, the dot density being a qualitative measure of the concentration of the element of interest. As may be seen in Figure 6.10, the quality of such dot maps is not very good, primarily because the counting statistics are very poor even for exposures of more than 1000 s.

A great improvement is obtained by using *digital X-ray mapping* to obtain this type of information. The concept of digital scanning electron microscopy

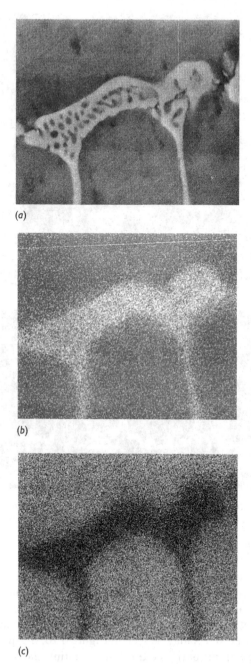

(a)

(b)

(c)

Figure 6.10 X-ray 'dot' distribution maps from the aluminium–copper alloy of Figure 6.9. (a) Secondary electron image. (b) Copper K$_\alpha$ map. (c) Aluminium K$_\alpha$ map. Despite the obvious presence of fine details within the copper-rich phase, the dot maps do not show it because of poor spatial resolution and poor counting statistics.

was introduced in section 5.10. In this method the beam is controlled by computer to move to a grid of points on the specimen, which is the digital equivalent of a raster. The beam remains at each of these points for a pre-set time, while an analysis is carried out. The data are stored in the computer memory and may be displayed on a grey level or colour scale. Each analysis point then becomes a pixel in the displayed image. One of the advantages of this method is that, with an EDS system, data from several elements may be collected at the same time by directing the counts from specified channels of the spectrum into different computer stores. An electron image may be acquired at the same time. Having acquired a digital X-ray map, the data may be subsequently manipulated as discussed in section 5.10. An example of digital X-ray mapping, in which X-rays from four elements were acquired simultaneously is shown in Figure 6.11. Using the same control technology, digital linescans may also be obtained.

In obtaining an X-ray map from a solid specimen, it must be borne in mind that the spatial resolution is limited to around 1 μm, and there is no point in working at high magnifications. For example, the X-ray maps of Figure 6.11 were obtained at a screen magnification of 200 (on a 10 cm × 10 cm screen) using a digital scan of 128 × 128 points which took 60 minutes. Each specimen pixel is thus approximately 4 μm × 4 μm, which is larger than the sampling volume, which it needs to be in order to avoid blurring of the image by overlap of data from adjacent pixels (see section 5.4).

Even so, the quality of an X-ray map does not compare with that of an electron image, and the reason for this, as we have mentioned before, lies in the poor counting statistics for X-rays as compared with electrons. This may clearly be seen if we analyse Figure 6.11. The total X-ray count rate was approximately 2000 cps and the beam current 10^{-9} A. The strongest peak was $Co\,K_\alpha$, for which 10 channels were counted. This amounts to around 15% of the counts for the whole spectrum. The count rate for $Co\,K_\alpha$ was therefore approximately 250 cps. The 16 384 pixels were scanned in a time of 60 minutes, and therefore each pixel represents an analysis time of 0·2 s. Therefore the average number of $Co\,K_\alpha$ counts per pixel is 50. Referring back to the discussion of signal noise in section 5.5.2, we find that the noise is therefore $\sqrt{50} = 7$, and, the minimum contrast level which can be distinguished in the image is given by equation 5.16, as 50%. Thus, considered as an image, the X-ray map is extremely poor. Compare this with the electron image of Figure 6.11(f), which was acquired in 100 s at the same beam current of 10^{-9} A. The number of backscattered electrons forming this image may be calculated from equation 5.18, making an adjustment to allow for a 128 × 128 pixel array, rather than the 1000 × 1000 array assumed in that equation, and this works out at over 3 million detected electrons per pixel. This discussion clearly shows that although X-ray maps are a very elegant way of displaying qualitative analytical information, they are not a practical proposition in most situations, and as stated above, a better and more rapid

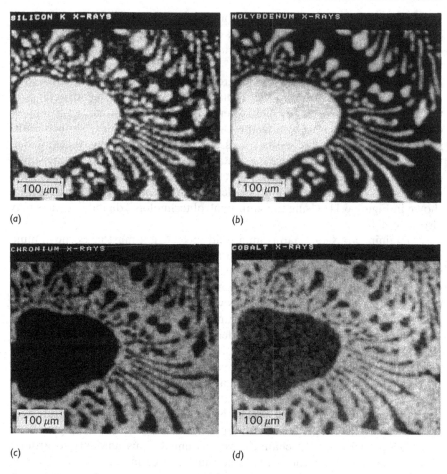

Figure 6.11 Digital X-ray maps obtained simultaneously from a polished specimen of a hard-facing alloy. The data were acquired in 60 minutes. (*a*) Silicon K$_\alpha$. (*b*) Molybdenum L$_\alpha$. (*c*) Chromium K$_\alpha$. (*d*) Cobalt K$_\alpha$. (*e*) Detail of (*d*) showing the individual pixels. (*f*) A digital backscattered image of the same area. This image was acquired in 100 s.

method of obtaining the same information is to use backscattered electron imaging in conjunction with X-ray spot analysis of selected areas.

6.3.3 Practical problems of qualitative analysis

When using any of the methods for obtaining analytical results which have just been discussed, it is necessary to bear in mind some practical problems which may arise with certain specimens.

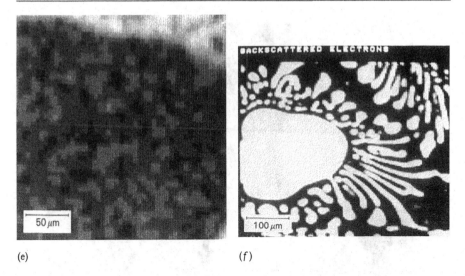

(e)　　　　　　　　　　　　　　(f)

Figure 6.11 continued.

Since the specimen is in a scanning electron microscope it must have a conducting surface (unless a low-voltage microscope is being used). Non-conducting specimens must therefore be coated before being examined. The coating, however, will absorb X-rays as they are emitted from the specimen and it will also emit its own characteristic X-rays. Consequently the coating should be as thin as possible, be of as low an atomic weight as possible, and it must not contain an element which might be of interest in the specimen. For these reasons, gold, which is generally used for SEM work, is less suitable as a coating material than carbon if the specimen is to be analysed.

Another important point is that as X-rays, unlike secondary electrons, travel in straight lines from the specimen to the detector, we will not detect X-rays from regions of the specimen which are not in the line of sight of the detector. Therefore on rough specimens we must be careful to interpret this as a topographical effect rather than one due to a variation of chemical composition across the specimen. An extreme example of this effect is shown in Figure 6.12(b), where an X-ray map of oxidized copper spheres, taken with oxygen K_α X-rays is shown. Comparison of (a) the secondary electron image and (b) the X-ray map clearly shows that the apparent variation of oxygen level is purely a topographic effect. For this reason it is preferable to work, where possible, with flat specimens for analysis.

Another important point arises from the fact that X-rays originate deep in the specimen, and therefore we may detect X-rays from regions which are not visible in the electron image. For example, as seen in Figure 6.13(a) and (b), we may find that X-rays from an unseen phase below the surface are being collected.

(a)

(b)

Figure 6.12 (a) Secondary electron micrograph and (b) oxygen K_α distribution map from oxidized copper spheres. The apparent variation of oxygen content across the spheres is a topographic effect in which X-rays from one side of the spheres cannot reach the crystal spectrometer.

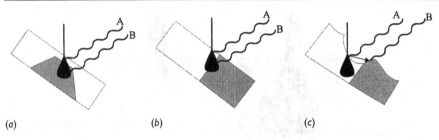

Figure 6.13 Three common situations in which a false impression of local composition can be obtained. (*a*) In a region apparently consisting entirely of A, a region of B below the surface emits B X-rays. (*b*) Near the boundary between phases A and B, both types of X-ray are excited although the electron beam appears to be on the B phase. (*c*) In a rough specimen, B X-rays may be excited by fluorescence from the A region.

Figure 6.13(*c*) shows another source of error which is particularly important for rough specimens. X-rays or high energy electrons emitted from the specimen may subsequently hit other areas of the specimen or even other parts of the microscope, and excite X-rays by fluorescence from these regions. This may of course produce a misleading analysis. WDS analysis is less prone to this problem than EDS analysis as the X-ray beam is well collimated by the spectrometer slits.

6.4 X-ray analysis of thin specimens in the TEM

We have seen in this chapter that X-ray analysis in the SEM or electron probe microanalyser is a very powerful analytical technique with much higher spatial resolution than conventional analytical methods. However, as discussed earlier, the spatial resolution is limited to the size of the sampling volume which is around 1 μm – much worse than the resolution of the electron microscope when used for electron imaging. There are many important features of a specimen which are much smaller than 1 μm, and there has consequently been a great deal of effort in developing methods of X-ray analysis with higher spatial resolution.

The problem may be overcome as shown schematically in Figure 6.14, by using thin specimens. Although the shape of the interaction volume depends somewhat on electron energy and the atomic number of the specimen, it is typically as shown in Figure 6.14(*a*), the lateral spread of the electron beam increasing with the depth of penetration. Therefore in a sufficiently thin specimen, as shown in Figure 6.14(*b*), the beam spread is much less. The shape of the truncated cone may be calculated, or determined from Monte Carlo simulation (see e.g. Hren *et al.*, 1979), and for thin specimens, the

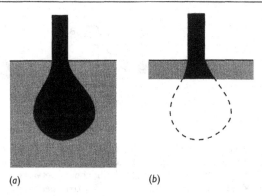

Figure 6.14 A comparison of the size of the region from which X-rays are excited in (*a*) a solid specimen, and (*b*) a thin specimen.

beam spread (in nm) is given by

$$B = 0.198(Z/E)(\rho/A)^{1/2} t^{3/2} \tag{6.3}$$

where Z = atomic number, A = atomic weight, E = electron energy (keV), ρ = density (g/cm^3), and t = thickness (nm).

Examples of beam spread for a range of materials are shown in Figure 6.15.

Many transmission electron microscopes are now fitted with EDS detectors so that chemical information as well as microstructural information may be obtained from thin specimens. One of the important consequences of using a very thin specimen is that the number of X-rays emitted is much smaller than for a bulk specimen, and it is this factor which often limits the spatial resolution of thin film analysis.

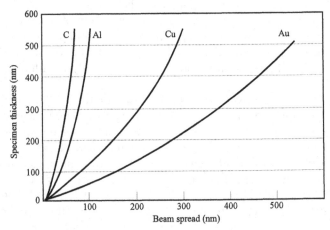

Figure 6.15 The effect of specimen composition and thickness on the lateral beam spread of 100 keV electrons in a thin specimen.

In order to maximize the collection efficiency we need to place as large a detector as possible, close to the specimen. However, it is not as easy to do this in a TEM as in an SEM, as the specimen is located within the objective lens. In most cases, use of an EDS detector requires the objective lens of the TEM to be fitted with pole-pieces with a wider gap than normal, with some loss of image resolution. Detectors are either mounted horizontally, or near horizontally through the polepiece gap, in which case the specimen must be tilted in order to allow X-rays to enter the detector, or else, if the geometry of the lens allows, an inclined detector is fitted. One of the advantages of a steeply inclined detector is that a specimen which is being examined in the TEM need not be specially tilted for analysis, and therefore any special diffracting conditions needed for imaging can be maintained. However, as larger-diameter detectors can usually be accommodated in the horizontal or near horizontal position, there are advantages for this geometry as well. It is not uncommon for microscopes to be fitted with detectors in both the positions, and the spectra of Figure 6.6 were acquired in such an instrument. The inclined detector had a standard beryllium window, and the horizontal detector was an ultra-thin window detector.

As discussed in Chapter 4, the electron beam in a TEM is normally not focused, but illuminates a large area of the specimen. In order to obtain X-ray analysis with high spatial resolution, the beam must only impinge on the area to be analysed, and therefore a fine focused probe must be used. There are two ways in which this is commonly done. STEM attachments for transmission microscopes have been available for a number of years. In order to fulfil their function, they need to be able to form fine electron probes, typically of diameters down to ~1 nm. Therefore, using such an attachment, X-ray analysis of a spot, line, or area of a specimen can be carried out in the same way as in an SEM, except that the area of interest is normally located by imaging in STEM. One of the problems of using this type of equipment at the highest spatial resolutions is that on changing from STEM mode to a stationary spot mode for analysis, it is difficult to know exactly whereabouts on the specimen the spot is located to an accuracy of better than perhaps 10 nm. Thus we can analyse a small volume of the specimen, but we are sometimes not sure exactly where it is!

In the latest generation of transmission electron microscopes, a very small probe (down to a diameter of 2 nm, or even less) can be formed when the microscope is in normal TEM imaging mode. Thus the area of interest can be located by TEM imaging, and the beam size can be reduced until it is smaller than the area to be analysed.*

---

* You should be careful to ask exactly what is meant by 'beam diameter'. Usually this refers to the full width at half maximum (FWHM) of a Gaussian distribution of electrons. However, this 'beam diameter' only contains 50% of the electrons.

Using either method, a spot size of less than 5 nm can readily be obtained, and therefore, if thin enough specimens are used (Figure 6.15), it should perhaps be possible to analyse areas of this diameter. However, the low current in such small probes (equation 5.13) combined with the small number of X-rays generated in such thin specimens, means that in a conventional TEM far too few X-rays to carry out a meaningful analysis can be collected. Therefore in a TEM or TEM/STEM instrument, the minimum spatial resolution for analysis, which will of course depend on the nature of the specimen and the concentration of the element being analysed, is obtained by optimizing the probe size and specimen thickness so as to obtain a sufficient signal. For a TEM/STEM instrument it is difficult to obtain a spatial resolution of much less than 10 nm for a typical specimen. Use of higher brightness electron guns, particularly the field emission gun in a dedicated STEM or FEGTEM instrument (Chapter 7) enables substantially better resolution to be obtained with suitable specimens.

In addition to the low count rate discussed above, there are one or two problems associated with TEM-based analysis which have to be weighed against the great advantages of a small selected area of analysis. In thin specimens, the 'surface' region may be a significant fraction of the total volume analysed, and it is therefore particularly important that the specimen is prepared in such a way that the surface chemistry is unaltered. Great care must be taken with electropolished metals, ion beam thinned materials and biological specimens to avoid this problem.

A further troublesome effect is the fluorescence or backscattered electron excitation of X-rays from the specimen holder and microscope column. The specimen in a TEM is much more closely surrounded by metal objects than is usually the case in the SEM, and mechanisms such as that illustrated in Figure 6.13(c) therefore occur frequently. These problems can be alleviated but not entirely cured by using a specimen and support grids made of carbon or beryllium. These then emit only very soft X-rays which are not normally detected, and therefore do not interfere with the analysis.

The applications of thin film analysis are widespread in the fields of materials science and biology, and many examples of the use of the technique will be found in the further reading section at the end of the book. Two fairly typical applications to materials are shown in Figures 6.16 and 6.17.

6.5 Quantitative analysis in an electron microscope

In our discussion of analysis up to this point we have been concerned only with the detection of elements in the specimen, i.e. qualitative analysis. There are many occasions when it would be highly desirable to know more about the specimen in a quantitative way.

For instance, three questions which are often of importance are: 'What is the size and shape of the region which has been analysed?', 'What is the smallest

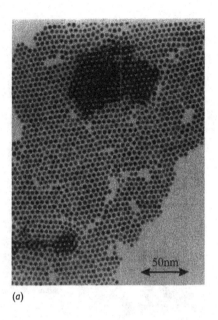

(a)

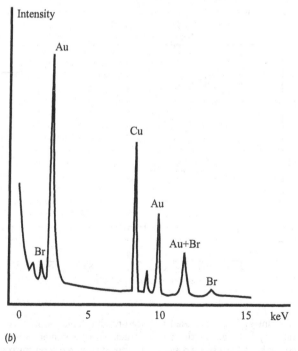

(b)

Figure 6.16 Analysis of ultrafine gold particles. (a) The EDS spectrum, (b) from a single 3 nm particle, shows that there is a small amount of bromine associated with each particle (C. J. Kiely and J. Fink, University of Liverpool).

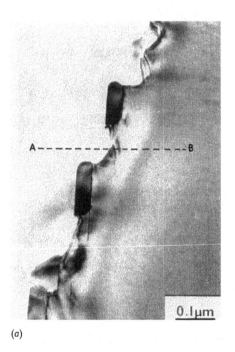

(a)

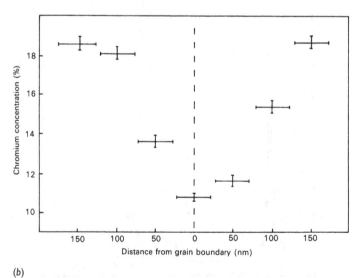

(b)

Figure 6.17 Thin film EDS analysis used to measure the depletion of chromium at a grain
boundary in an austenitic stainless steel. This phenomenon makes such an alloy
susceptible to stress corrosion. (a) Transmission electron micrograph showing a
grain boundary containing particles of chromium carbide. (b) Chromium concentrations in the matrix determined from spot analyses as the boundary is traversed
from A to B. (M. G. Lackey, Imperial College, London)

amount of element X which could be detected?' and 'How much of this element is present in this part of the specimen?' To find even an approximate answer to any of these questions requires a far deeper consideration of what happens inside the specimen as the X-rays are first generated and then partially absorbed again, than we have so far had space for. All we can hope to achieve in this section is some understanding of why these questions are difficult to answer precisely and what sort of order of magnitude the answers are likely to have.

Let us take the *volume analysed* first. Obviously, it is that region from which the X-rays which reach the detector are emitted. As we have seen at the beginning of this chapter this must depend on where the X-rays are generated (i.e. how far the electron beam penetrates the sample) and how strongly the X-rays are absorbed by the specimen on the way out. The volume is therefore going to depend on the electron beam energy, the average atomic weight of the sampling volume, the wavelength of the characteristic X-rays being studied, the absorption coefficients of the specimen for these X-rays and the angle of incidence of the electrons on the surface, to mention only the five most obvious factors. Another confusing point is that, of the X-rays emitted, more will probably come from near the surface than will be able to escape from deeper inside the specimen. The analysis is therefore not uniformly representative of the whole sampling volume. Since we would have to make a large number of not very accurate assumptions in order to calculate theoretically the volume analysed, this is not normally attempted. The analyst has to be content with knowing that the analysis from a bulk specimen is likely to come from a volume of about $1 \mu m \times 1 \mu m \times 1 \mu m$, or, as seen from Figure 6.15, much less from a thin specimen. It is up to the analyst to select the appropriate technique and operating conditions to ensure that the sampling volume is smaller than the region of interest.

6.5.1 The limit of detectability

The minimum limit of detection for any element in a specimen is of particular importance when small concentrations are suspected. It is not of much use to be told 'iron cannot be detected in your sample', unless this enables us to put an upper limit on the amount which could be present. The microanalyst therefore never gives this answer, but instead says that 'there must be less than (say) 0·1% of iron in your specimen'. X-ray counts, like other signals in the SEM, arrive at random, and we can only detect a peak in the spectrum if it is discernible from the background. The analysis of this situation is therefore analogous to the discussions in section 5.5.2 of the minimum discernible contrast level in an image.

If the mean count level of the background is $\bar{N}$, then the noise, or variation about this mean is $\sqrt{N}$. It is usually accepted that a peak can be recognized if it rises above the background level by at least $2\sqrt{N}$. If the count rate in the background is b counts per second, and if the counting time is t, then the background level is bt, and the smallest detectable peak is $2\sqrt{(bt)}$ above the background. What concentration of an element in the specimen would give this peak? To answer this we must compare our peak height with the peak height $(P - B)$ given by a standard specimen consisting of the pure element. To a first approximation the minimum detectable concentration (MDC) is given by;

$$\text{MDC} = \frac{100(P - B)_{\text{specimen}}}{(P - B)_{\text{pure standard}}} = \frac{200\sqrt{bt}}{pt - bt} \text{ weight } \%$$

where p is the peak count rate on the pure element standard. Simplifying this slightly we get

$$\text{MDC} = \frac{200\sqrt{b}}{(p - b)\sqrt{t}} \% \tag{6.4}$$

Clearly, if we wish to lower the MDC we can, according to equation 6.4, reduce b, increase p, or increase t. The easiest of these is to increase the time of analysis. A comparison of Figures 6.6(b) and 6.6(c) which were obtained in times of 50 s and 300 s respectively clearly shows how a longer counting time allows smaller peaks to be identified. However, it becomes impractical to count for longer than about 1000 s, not only because this would make a series of analyses very time consuming, but also because the beam intensity does not stay constant indefinitely, even on the best instruments, and if it varies perceptibly it will no longer be valid to compare the counts from the specimen with those from the standard, since they cannot be obtained under the same conditions.

The other ways of improving the MDC really depend on the design and functioning of the analysis and detector system. EDS systems have poorer peak-to-background ratios than WDS systems, and both systems have worse peak-to-background ratios for the lighter elements. For all but the lighter elements, the MDC using an EDS system is around 0·1%, and that using WDS, an order of magnitude lower. However, the precise figure will depend on the atomic weight of the trace element compared to that of the specimen, because, as will be discussed in the next section, the soft X-rays from a light element will be strongly absorbed by a matrix of high atomic weight.

6.5.2 Quantitative analysis of bulk specimens

The principle of quantitative analysis is simple, and we have implicitly made use of it in discussing the minimum detection limit. To estimate the amount of an element present we determine the number of characteristic X-ray counts from the specimen in a fixed time interval, N_{spec}, and compare this figure with the number arriving from a standard of known composition in a similar time,

N_{std}. The concentration of this element in the specimen, C_{spec}, should then be given by

$$C_{spec} = \frac{N_{spec}}{N_{std}} \times C_{std} = k \times C_{std} \tag{6.5}$$

where C_{std} is the accurately known concentration of this element in the standard, and each value of N is a peak count minus a background count. The background in an EDS spectrum varies continuously across the spectrum as may be seen in Figure 6.6, and accurate determination of the background level for any peak is not simple, although various modelling and filtering routines have been developed to accomplish this task.

As the two counts N_{spec} and N_{std} have to be collected at different times, we must be certain that the analysis conditions have not changed between analysis of the specimen and the standard. This calls for a great deal of care, both in designing electron microprobe analysers and using them, in order to ensure that, in particular, the electron beam current remains constant for long periods. Even if this problem is overcome, we find that equation 6.5 is totally inadequate for the calculation of concentrations.

Most of the complicating factors arise because, inevitably, the specimen is not the same as the standard. In the most usual cases a specimen containing several elements is compared with a series of standards, each of which is a pure element, and the specimen is therefore likely to differ from each standard in its density and in the average atomic weight of its constituent atoms. The consequence of these differences is that equation 6.4 may need correcting for some of, or, in the worst cases, all of the three factors known as *atomic number effect* (*Z*), *absorption* (*A*) and *fluorescence* (*F*). Use of these corrections is commonly known as the ZAF technique. Although analytical equipment suitable for quantitative analysis is always linked to computers equipped with the programs to rapidly carry out the ZAF corrections, the analyst needs to understand their principle in order to ensure that the assumptions made in the ZAF corrections are appropriate to the analysis being carried out.

Before considering the nature of these corrections, we will discuss the method by which the ZAF technique works. X-ray counts from an element are measured from both the specimen and the standard (which we will assume to be the pure element), and their ratio, k, is calculated. C_{spec} is given by

$$C_{spec} = k \times Z \times A \times F \tag{6.6}$$

As we will find later in this section, the values of Z, A and F depend on factors such as the mean atomic weight of the specimen and calculation of them therefore requires a prior knowledge of the specimen composition. The only way to proceed is to make a rough estimate of C_{spec} usually by taking it as k, (i.e. $Z = A = F = 1$), and to use this value to calculate approximate values of Z, A and F. These values are then inserted into equation 6.5 and used to calculate a more accurate value of C_{spec}, and this iterative calculation is repeated until a constant value of C_{spec} results. No more than three or four iterations are normally required to obtain the result. None of the three correction factors can be calculated precisely from first principles, but over the past 40 years, approximate solutions have been obtained and continuously refined.

The *atomic number correction*, Z, is concerned with the efficiency with which an element generates X-rays, and this depends on two factors; (a) how far the electrons penetrate the specimen before they lose too much energy to excite further X-rays (see Figure 6.2), and (b) how many electrons are backscattered without exciting any X-rays (see Figure 5.8(a)). When the mean atomic number of the specimen differs considerably from that of the standard, the count rate for an element will not be linearly proportional to the amount present, and Z will therefore not be equal to 1. For analysis of a heavy element in a light element matrix $Z > 1$, and for a light element in a heavy element matrix $Z < 1$. As an example of the magnitude of the effect, for sulphur (atomic number $= 16$) in a stainless steel (mean atomic number $= 27$), Z is 0·87. This is an extreme example, and the effect is often much smaller. In determining the vanadium concentration (atomic number $= 23$) in the same steel, the correction factor Z is 0·99.

The *absorption correction*, A, is also very important. As can be seen from Figure 6.2, many of the X-rays emerging from the specimen will have travelled a considerable distance within the solid and their intensity will have been reduced by absorption. The amount of the absorption depends very strongly on the elements in the specimen, through their mass absorption coefficient, μ. It is quite likely that the specimen and the standard will have different mean absorption coefficients and therefore, even if X-rays were generated in exactly the same interaction volume in the standard and the specimen, the amount of absorption would differ. The magnitude of this correction can be quite large, especially where soft X-rays (e.g. the K lines of light elements, see Table 6.1) are emitted from specimens containing heavier elements. A severe example is the measurement of aluminium in a glass containing alumina (Al_2O_3), silica (SiO_2) and lime (CaO). The factor A in equation 6.6 is then 1·86. The absorption correction is often the most significant effect; for example, even the iron analysis in a stainless steel might need an 8% correction $(A = 1·08)$. The accuracy of any correction procedure depends on the values of μ which are used. Good experimental values are not available for all elements and X-ray lines (particularly for light elements) and this may be a limiting factor in many corrections.

The final factor to be considered is *fluorescence*. From its very nature (see section 6.1) fluorescence cannot occur within a pure elemental standard. However, in a specimen containing several elements it must be considered. Fortunately fluorescence is a very 'inefficient' process and only a very small proportion of high energy X-rays excite lower energy fluorescent radiation. However, when elements of nearby atomic number are present, as tends to happen for example in steels (Cr $= 24$, Mn $= 25$, Fe $= 26$, Ni $= 28$)

the fluorescence effect, which gives rise to more of the lower energy X-rays than would be expected, can be important. One of the worst cases is that of chromium in steels where a correction of 15% ($F = 0.85$) may be needed.

The application of computer correction procedures to good results from the analysis of solid specimens using well characterized standards should enable concentrations to be calculated to about $\pm 2\%$ with a WDS and $\pm 6\%$ with an EDS. However, these results can only be obtained by extremely careful experimentation and where there are large amounts of the element present (say about 10%). Thus, although it might be possible to measure the concentration of copper in a brass as $70.0 \pm 1.4\%$, the carbon content of a steel could scarcely be calculated and quoted better than $0.3 \pm 0.2\%$ because of the low concentration and the lack of accurate values for absorption coefficients of the 'soft' carbon K_α X-rays.

One of the most important parameters in the correction equations is the X-ray take-off angle (ϕ in Figure 6.7) which determines the X-ray path lengths in the specimen. It is extremely important that this should be the same on both specimen and standard and, indeed, at all points of the specimen which are to be analysed. Consequently any specimen for quantitative analysis must be flat on the scale of the electron beam diameter. It is usual to polish specimens immediately before analysis with an abrasive of $1\,\mu m$ or $0.25\,\mu m$ particle size so that only very fine undulations are present. A flat specimen is also virtually essential for WDS analysis since, as Figure 6.7 shows, the specimen must lie exactly on the Rowland circle for efficient detection of the X-rays. A rough specimen could not be moved sideways at all without re-aligning it on this circle which would be a great inconvenience. It is one of the apparently great advantages of the EDS detector that it is extremely insensitive to the position of the specimen and, therefore, rough specimens can easily be 'analysed'. However if a quantitative analysis is required the take-off angle must be known and, therefore, a polished specimen is again necessary. It is worth bearing in mind that even a qualitative analysis from a point on the specimen using an EDS can be misleading because of the effects shown in Figure 6.13(c).

Although the preparation of the specimen is a key factor in quantitative analysis there are also problems associated with the standards. For many elements, particularly the metals, it is extremely easy to prepare a polished sample of the pure element. However, many of the other elements exist as gases, liquids or very reactive solids, and standards for these elements must therefore be in the form of compounds. Care must be taken to choose compound standards which are homogeneous, stoichiometric and fully dense. In order to overcome these limitations much work has been done in the 1980s and 1990s to develop 'virtual standards'. These are computer-based spectra which can be used, under certain carefully controlled conditions, to replace standard spectra. Since virtual standard spectra do not need to be collected at the time of analysis their use speeds up the analysis process considerably as well as obviating the need for complicated or not very stable real standards.

Since the major emphasis in this section has been on the quantitative analysis of metallic or mineral specimens, which can generally be polished flat without damage, it may be useful to summarize some of the problems involved in applying these techniques to biological material. The majority of biological samples are unsuitable for accurate quantitative analysis as bulk samples because they do not fulfil three of the essential criteria discussed earlier in this section:

(a) They are rarely stable under electron bombardment, frequently changing their volume and their chemical nature.
(b) They are often not homogeneous in composition over the interaction volume, which in low atomic weight material may be as large as $10\,\mu m^3$.
(c) They do not normally maintain a flat surface inside the microscope.

By using standards which have chemical compositions similar to the specimen, some success may be achieved in favourable circumstances. However, the best chance of successful quantitative analysis of biological materials is in thin sections, using the methods described in the next section.

6.5.3 Quantitative analysis of thin specimens

We saw in section 6.4 that there was a significant gain in the spatial resolution of analysis if very thin specimens were used. Under certain circumstances quantitative analysis of such thin sections may also be easier to carry out than for bulk samples. This is because the path lengths of the X-rays in the specimen are so short (Figure 6.14) that the absorption and fluorescence corrections necessary for bulk samples may be neglected. It has been shown by Cliff and Lorimer that for very thin samples

$$\frac{C_A}{C_B} = k_{AB}\frac{N_A}{N_B} \tag{6.7}$$

where N_A and N_B are the measured characteristic X-ray intensities, and C_A and C_B are the weight fractions of any two elements A and B in the specimen. The scaling factor k_{AB} is dependent on the two elements, the operating conditions and the detector response. Although it is possible to calculate k_{AB}, it is more usual to determine it experimentally. In order to avoid determining k for all combinations of elements, it is more convenient to measure k for all the required elements in combination with a single element, typically silicon, i.e. k_{ASi}, using alloys or minerals of known composition. When this set of constants has been evaluated, then k_{AB} in equation 6.6 may be replaced by k_{ASi}/k_{BSi}.

As discussed in section 6.4, one of the main problems in thin-film X-ray analysis is the detection of spurious X-rays, and for quantitative analysis it is important that their level is reduced as much as possible. Having obtained spectra, methods of determining the characteristic X-ray intensities are similar to those used for bulk specimens.

The limit of detectability of element A in a matrix of element B is given by Goldstein as

$$C_A = \frac{3(2I_c)^{1/2}}{I_B} \cdot k_{AB} \cdot C_B \tag{6.8}$$

where I_B is the intensity from element B and I_c is the continuum background for element A. Equation 6.8 shows that the detectability decreases (i.e. gets better) as the count rate or counting time increases. As discussed in section 6.4, an increased count rate in a given instrument can be achieved only with a thicker specimen or a probe of larger diameter. Increasing these factors will lead to a deterioration of spatial resolution, and if the thickness is increased too much then absorption and fluorescence corrections are needed and the simple ratio method for analysis (equation 6.7) breaks down. In practice, detectability limits are typically of the order of 0·1–0·5 wt%. The thickness limit for which equation 6.7 is valid depends on the element being analysed, and the atomic weight of the specimen. Typically it is of the order of 50–100 nm.

For thicker specimens, absorption corrections may be made. The absorption correction depends critically on the path length in the thin film over which absorption occurs. This necessitates a knowledge of not only the local specimen thickness, but also the specimen shape and tilt, and these parameters are often difficult to determine accurately. As is the case for bulk specimens, use of the correction requires an iterative technique.

Despite the problems of thin film analysis, it is an extremely powerful microanalytical tool, and is widely used for phase identification, and the determination of elemental segregation or depletion on a fine scale as shown in Figure 6.17.

Although bulk microprobe analysis is often used to determine the composition of phases in a ceramic or metallic alloy, in many cases the distribution of the phases is on such a fine scale that the 1 μm spatial resolution of the technique is inadequate. In these circumstances thin film analysis is extremely useful and has made many valuable contributions to the study of phase transformations and phase equilibria.

6.6 Electron energy loss spectroscopy (EELS)

In section 2.7 we discussed four types of inelastic scattering. If, during its passage through a thin specimen, an electron undergoes any of these processes it will suffer a loss of energy. If the energy loss is greater than about 1 eV then it is relatively simple to measure it with a magnetic spectrometer. This forms the basis of a powerful way of investigating the nature of the specimen in a TEM.

6.6.1 The spectrometer

An electron spectrometer must obviously be mounted after the specimen and is usually the final component of an analytical microscope. The principle of the instrument is that a magnetic field is used to deflect all the electrons through about 90 degrees. The more energetic electrons will be deflected slightly less and the beam will therefore be dispersed into a spectrum of energies. Figure 6.18 illustrates this. There are two ways of detecting the spectrum. If a single detector is available then the spectrum can be scanned across a slit, by varying the strength of the magnetic field, so that each energy is detected in turn. This is known as a *serial spectrometer* and is shown in Figure 6.18(*a*). If however a multiple 'position-sensitive' detector is used, then the whole spectrum can be collected at once. This arrangement is called a *parallel spectrometer* and the technique is known as PEELS (parallel electron energy loss spectrometry) (Figure 6.18(*b*)). Most early spectrometers were of the serial type. However, parallel detectors are now widespread since they are intrinsically more efficient. Assume that we wish to count the number of electrons with one thousand different energies in order to build up a spectrum consisting of one thousand points. A serial spectrometer would spend only one thousandth of its time counting at any one energy whereas a parallel spectrometer counts all

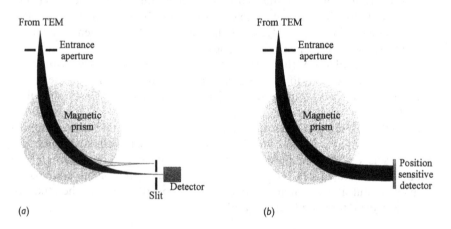

(a) (b)

Figure 6.18 Electron energy-loss spectrometers: After passing through the specimen (and the projector lenses of the microscope) electrons enter the spectrometer and are deflected by the magnetic field. In a serial spectrometer (*a*) a slit is used to ensure that only electrons of a single energy loss hit the detector. The field is then varied so that each energy in turn is detected. In the parallel spectrometer (*b*) a position sensitive detector is used to collect electrons of all energies in parallel. No scanning of the magnetic field is then necessary.

energies all of the time. Inefficiencies in the parallel detector itself mean that the advantage is not quite one thousand to one in its favour but nevertheless parallel detection systems are currently the dominant spectrometers.

6.6.2 The spectrum – qualitative analysis

An EEL (or PEEL) spectrum is conventionally considered to consist of three regions, shown in Figure 6.19. Those electrons (usually the majority) which have suffered negligible inelastic scattering contribute to the *zero loss peak*. The *low loss region*, containing electrons which have lost up to about 50 eV, arises largely from plasmon scattering. There may be several plasmon loss peaks. The first, at the lowest energy loss, corresponds to those electrons which have excited one plasmon in their passage through the specimen. Second and higher plasmon peaks arise from those electrons which have excited two or more plasmons. Equation 2.7 can be used to calculate the relative intensities of plasmon peaks. In most spectra from reasonably thin specimens the first plasmon peak will be the most intense, with subsequent peaks getting progressively weaker. If the fourth plasmon peak can be seen the specimen must be quite thick and will not be very useful either for imaging or analysis.

Plasmon peaks are not very useful for analysis because the energies at which they occur are similar for many materials. For analytical purposes it is more useful to study the *characteristic edges* at higher energy losses. The number of electrons which suffer high energy losses is small. However the energies of the edges which correspond to inner shell excitation (section 2.7.4) are just as

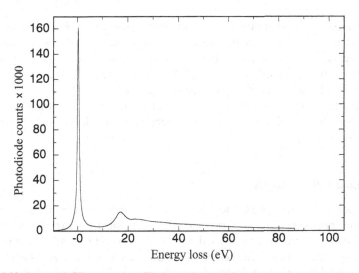

Figure 6.19 A typical EEL spectrum. The zero loss and low loss and regions are evident at energy losses of 0 and around 20 eV, but the characteristic edges are too small to be seen on this scale.

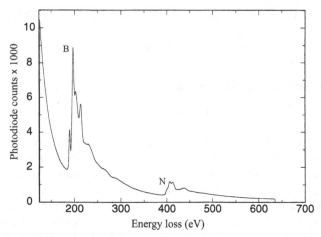

Figure 6.20 Part of an EEL spectrum from boron nitride (BN), showing the boron and nitrogen K edges.

characteristic of the element as are X-ray peaks. It is these edges which are routinely used for EELS analysis.

Qualitative analysis is carried out by determining the energies of any edges which can be seen, usually between 100 eV and 2000 eV, and comparing them with tabulated values for the elements. Thus in Figure 6.20 the edge at 188 eV arises from the K transition in boron while the edge at 399 eV corresponds to nitrogen.

Because they arise from the same electronic transitions, EELS edge energies will be close to, but not exactly the same as, the energies of characteristic X-rays. In general of course EELS (absorption) edge energies should be slightly higher than the corresponding X-ray (emission) peaks. One of the great strengths of EELS is that it can be used to detect edges from elements whose X-rays are too soft to be counted. Thus for example it is possible to detect EELS edges from helium, lithium and beryllium which are otherwise very difficult to analyse.

6.6.3 Quantitative analysis

In principle it is relatively easy to deduce from an EEL spectrum the composition of the specimen. The 'size' of each edge is proportional to the number of atoms of that element in the analysed region of the specimen. If we can identify an edge from each element in the specimen, and measure the 'size' of each edge, then we should be able to calculate the composition. This is indeed the basis of quantitative analysis, but there are a number of reasons why it is not quite as simple as suggested in the above three sentences.

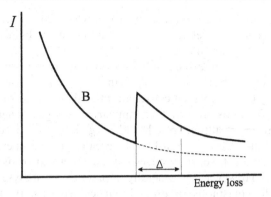

Figure 6.21 A schematic EELS edge showing the energy window Δ over which the integrated edge intensity is measured. The background is fitted to the region B and then extrapolated as shown by the dashed line.

We must first consider the probability that electrons of the energy being used will excite an atom. This is deduced from the cross-section for inner shell excitation, defined in section 2.5. The cross-section for excitation of a particular shell (we will use the K shell as an example) depends on the element, on the angular range of electrons which are accepted into the spectrometer (β) and on the energy window (Δ) which is to be considered (shown in Figure 6.21).

The cross-section decreases for elements of high atomic number (equation 2.9), but fairly obviously increases as β or Δ are increased. A spectrometer of fixed β will show light element edges far more intense than those from even medium atomic weight elements. This is why, in Figure 6.20, the boron edge is so much bigger than the nitrogen edge. It does not mean that the specimen contained more boron than nitrogen. In fact there should be equal numbers of boron and nitrogen atoms in boron nitride.

Inner shell cross-sections can be calculated from theory and this approach is generally used to determine a 'weighting factor' for each edge. It then remains to determine what is the 'size' of the edge. This is defined as the area of the spectrum *above the background* within the energy window Δ. The background falls very rapidly with energy, as all the spectra in this section show, so extrapolating the background beneath each edge is difficult. The most common approach is to fit the curve just below the edge (i.e. at slightly lower energy loss) to an equation of the form $A e^{-r}$, and then to use the measured values of A and r to extrapolate the background beneath the edge (Figure 6.21).

Since edges from medium atomic mass elements are so small the EELS technique is most powerful for analysing light elements. If not all the elements in the specimen are to be determined using EELS then it is not possible to calculate the composition completely using the technique outlined above. An alternative approach is to compare the size of each edge with the size of the zero loss peak over the same energy window Δ. This permits the calculation of the absolute number of atoms which contributed to the edge. Using this approach the relative atomic abundances of light elements in the specimen are easily determined.

One of the major limitations of EELS analysis is that each edge will be convoluted with (i.e. broadened in the same way as) the low-loss region. In other words any electron which causes an inner shell excitation can *also* excite one or more plasmons. It will then appear at a higher energy loss in the resultant spectrum, resulting in a broadening of the ideal edge shape. The magnitude of this effect obviously gets greater as the specimen thickness increases. The cleanest, sharpest edges will be generated from very thin areas of the specimen. However it is generally impractical to analyse a region which is so thin that few plasmons are excited, particularly since the total signal (i.e. number of counts in the spectrum) will then be very small and statistical inaccuracies will be large. In most analyses a compromise is therefore struck between signal strength and edge blurring. However equation 2.7 shows that if the specimen thickness is equal to the mean free path for plasmon excitation, 18% of electrons will excite two plasmons, so the second plasmon peak will be quite substantial. Even at a thickness of half the mean free path the second plasmon peak will contain 7% of the electrons. Since plasmon mean free paths are about 100 nm, the analysed region clearly needs ideally to be substantially thinner than this.

6.6.4 Extra information from EEL spectra

An EEL spectrometer is not only useful for light element analysis. It can also be used to deduce more information about the specimen. We will give just two examples of this. Since plasmon scattering is well described by the Poisson equation (2.7) if a specimen is thick enough to give rise to a substantial first plasmon peak we can take the ratio

$$\frac{p(1)}{p(0)} = \frac{(t/\lambda)\exp(-t/\lambda)}{\exp(-t/\lambda)} = \frac{t}{\lambda}$$

From the experimental ratio of the first plasmon peak intensity to the zero loss peak we can therefore calculate the thickness in units of the plasmon mean free path. Since the mean free path can be looked up or calibrated once with a specimen of known thickness, this is a rapid technique for determining local thickness.

A second type of information is present in the *fine structure* before and after each edge. The structure near the edge is known as energy-loss near edge structure (ELNES) and can be quite different for an element in different compounds. For example, Figure 6.22 shows edges for Al in metallic aluminium and aluminium oxide. The shapes of the edges are quite distinct and enable us to determine not just that Al is present but that it is in the form of aluminium oxide. The small variations in intensity after the

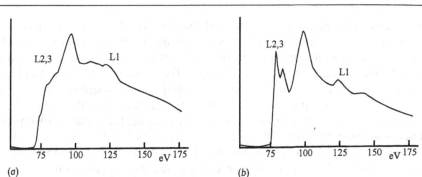

(a) (b)

Figure 6.22 Aluminium L2,3 edges in EELS spectra from (a) metallic aluminium and (b) aluminium oxide Al_2O_3. Notice the major differences in the region before the main edge. Both edges have had their background removed.

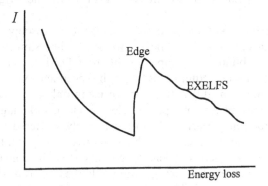

Figure 6.23 A single EELS edge showing extended fine structure (EXELFS).

edge (EXELFS = extended energy-loss fine structure, shown schematically in Figure 6.23) carry information about the local environment of the atoms contributing to the edge, that is about their bonding and co-ordination. The variation in signal is small and the signal itself needs to be strong before statistically reliable conclusions can be drawn. However parallel spectrometers are now widely available and this technique will be increasingly useful.

Elemental mapping can be carried out for thin specimens in much the same way as has been shown for solid samples in Figures 6.10 and 6.12. The X-ray (EDS) techniques discussed in section 6.5.3 can be used to form X-ray elemental maps from thin specimens. However it should always be remembered that the total X-ray signal from a thin specimen is substantially lower than that from the larger volume of a thick sample so that counting statistics will be significantly poorer and an X-ray map from a thin specimen will usually look very 'noisy'.

Elemental mapping can also be done using EELS. However the large background beneath most energy loss edges (e.g. see Figure 6.20) means that the contrast in such maps is rather low, and therefore the detection limit tends to be rather large. Put another way, even if there was no boron present at 188 eV in Figure 6.20 the apparent signal from that part of the spectrum (all resulting from the background) would be quite large. The solution to this problem is to collect a full spectrum at each point of the map and subsequently to process each spectrum to subtract the background before displaying the image. This technique is rather heavy on computer power because, for a 1000×1000 pixel image you need to store 10^6 spectra of about 1000 data points and then process the 10^6 spectra to extract the elemental intensity or concentration at each point. This is rapidly becoming feasible as computing power becomes cheaper and it offers the huge advantage that you could go back to the original data long after the specimen had been taken out of the microscope and reprocess the spectra to look at the distribution of another element, which did not appear to be of interest at the time the original data was collected.

A further exciting possibility is to produce, at the microscope, energy-filtered images. This can currently be done in two ways. The simplest conceptually (although it means building a completely different TEM) is to insert an 'omega filter' into the microscope column just after the specimen so that the remaining lenses form the image or diffraction pattern using only electrons of a selected energy. This enables the microscopist to select an elemental image or a plasmon image or a diffraction pattern with no inelastically scattered background for instance. The alternative method, which involves less modification to the microscope, is to use a series of lenses and slits after the conventional EELS spectrometer to reconstruct the image using only the electrons which have been selected by the spectrometer. The most familiar version of this is known as a GIF, for Gatan Imaging Filter. Both techniques are likely to become more widespread.

6.7 A brief comparison of techniques

In this chapter we have dealt with the most commonly available analytical techniques associated with SEM and TEM. We have not been concerned with Auger electron spectroscopy, which has a great deal to offer but which is treated in Chapter 7. We have shown that X-ray analysis of solid samples is now a standard quantitative technique, extending to all elements down to boron in the periodic table. For the analysis of thin specimens there are two techniques, EDX and EELS, of which EDX is currently slightly more versatile in that it can more easily deal with a range of elements throughout the periodic table (boron upwards). However EELS can be used to analyse very light elements (helium upwards) and has a number of other uses in addition to chemical analysis. Elemental mapping is available using a variety of techniques.

6.8 Questions

1 You have available an analytical microscope with an EDX detector and maximum electron energy of 20 kV. Using the data in Table 6.1 suggest which line you would use to analyse for P, Mo, Cr and Os.

2 The energy required to excite an electron-hole pair in a silicon EDS detector is 3·8 eV. Calculate the number of pairs likely to be generated by a Be K_α X-ray and a Zr K_α X-ray and give two reasons why the analysis of beryllium is likely to be less accurate than analysis of zirconium.

3 During EDX analysis using K lines, which elements might be spuriously detected
(a) in Cr because of an escape peak
(b) in Al because of a sum peak?
Can you find, from Table 6.1, any other escape or sum peaks which might easily be misinterpreted?

4 In a crystal spectrometer (WDX) what is the approximate interplanar spacing which is needed tin the curved crystal (Figure 6.7) to detect and analyse:
(a) the Fe K line
(b) the Ti L line?
What sort of crystal is suitable for analyses using the Ti L line?

5 Beam spreading in a thin specimen is given by equation 6.3. A 100 keV beam of diameter 5 nm is used to analyse small particles in a copper specimen of thickness 200 nm. What is the effective diameter of the beam when it emerges from the specimen? How would you expect the position of the particle in the specimen (i.e. at the top, middle or bottom) to affect (a) a single spot analysis with the beam stationary, and (b) an elemental map formed with a scanned beam. The density of copper is $8·96\,\mathrm{g\,cm^{-3}}$ and its atomic mass is 63·5.

6 An aluminium–copper sample is being analysed using EDX. The pure Al standard gives a count rate of 2000 cps while the background intensity away from the Al K peak is 25 cps. Estimate the minimum detectable concentration of copper in this sample using an analysis time of (a) 0·15; (b) 10 s, and (c) 10 min.

7 Williams and Carter state as a guideline for EELS that 'if the intensity in the first plasmon peak is greater than one-tenth of the zero-loss intensity, then your specimen is too thick for analysis'. If you are analysing an aluminium-based material with a plasmon mean free path of 120 nm, what is the maximum useful thickness of your specimen? Assume that all energy loss is accounted for by plasmon scattering (not too bad an assumption in most cases).

Chapter 7

Electron microscopy and other techniques

We have described in earlier chapters the basic techniques of scanning and transmission microscopy and diffraction. There are now many other techniques available for characterizing materials, each of which has some special advantage. In this final chapter an attempt is made to outline some of the most important techniques which can be used to complement electron microscopy and to point out the areas of overlap between the techniques. We also try to indicate how electron microscopy is developing, often in response to developments in other areas. Most modern investigations, particularly in the materials sciences, exploit more than one technique and it is important to know what each can offer, even if one cannot be expert in all of them.

7.1 Complementary imaging techniques

All imaging techniques use one of only two basic principles: Either the whole field of view is illuminated and lenses are used to form a whole field image (i.e. the pixels are examined in parallel), or the field of view is interrogated and the image is displayed point by point, usually by scanning (i.e. the pixels are examined in series). In this book we have concentrated on one main example of each of the types, the TEM (parallel imaging) and the SEM (serial imaging). In the remainder of this final chapter we will describe (rather briefly) a number of other microscopes – you should bear in mind that all of them conform to one or other of the two basic types. Nearly all of the newer techniques use serial imaging. The differences among them lie in two main areas; the nature of the illumination (light, electrons, ions or whatever), and the nature of the signal which is used to form the image or extract analytical information (light, electrons, ions, X-rays or whatever). The details of how the image is formed, that is how the scanning is achieved or how the lenses work, are less important at this stage than a clear understanding of the nature of the illumination and the imaging signal, since these tell us in principle the useful range of the technique.

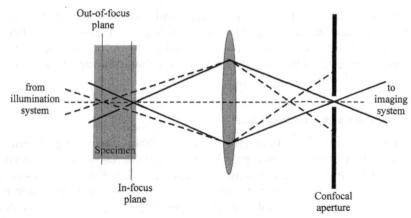

Figure 7.1 A schematic diagram of the key elements of a confocal microscope.

7.1.1 Confocal light microscopy

A major limitation of conventional light microscopy is that its depth of field is usually rather small. Anyone who has used a light microscope will be familiar with the narrow range in which the specimen appears in focus. For the biologist and others using transparent specimens the problem is compounded by the fact that one region of the specimen is in focus while others, which may be just as bright, are out of focus and appear blurred.

The very simple way round this problem, the use of 'confocal' optical systems, was discovered in the 1950s but we had to wait until the 1990s for the technology of lasers, computers and control theory to come together to produce user-friendly confocal microscopes. These are now spreading rapidly into many fields of science.

The basic confocal principle is very simple and is shown in Figure 7.1. Small apertures are used to ensure that only the light from the in-focus region of the specimen reaches the imaging part of the optical system. Thus the solid (in focus) rays form the image while the dotted (out of focus) rays are largely stopped by the aperture and contribute almost nothing to the image. The difficulty which accounted for the thirty year delay in putting this simple idea into practice is that only one point of the specimen can be genuinely confocal at a time. There therefore has to be a scanning system to illuminate one point of the specimen at a time. Since users are very interested in transparent objects the focused spot has to be scanned in three dimensions (i.e. through the thickness of the specimen as well as from side to side) and the 3D image has to be stored in a computer for later viewing. This is now easily done, using lasers and computer-controlled positioning devices, and confocal microscopes are readily available commercially. There are still a number of different ways of actually performing the necessary scanning, and a variety of imaging modes including fluorescence, but we will not deal with them here.

Sheppard and Shotton (1997) give a more detailed introduction to these techniques. Confocal microscopes are now widely used by biologists but also by materials scientists who need to study the topography of non-flat specimens at the resolution offered by the light microscope without having to put them in the vacuum of an SEM.

7.1.2 Scanned probe microscopies

For almost twenty years, from 1965 to the mid 1980s, the imaging of surfaces at a resolution better than that provided by a light microscope was dominated by the SEM. Alternative methods, including reflection electron microscopy, were available for the whole of that time but they were on the whole not refined into useful high resolution techniques. However since the invention of the scanning tunnelling microscope there has been an explosion in the area of scanned probe techniques. The atomic force microscope (AFM) has become almost as ubiquitous as the SEM and many other variants of the basic technique have been developed. In this section we will describe the basic principle of the techniques and give a brief review of the range now available.

Scanning tunnelling microscopy

The scanning technique which opened up the new range of techniques is scanning tunnelling microscopy (STM). Since the publication of the first edition of this book, the technique has been invented, developed, refined and marketed to many laboratories around the world. Its inventors won (jointly with one of the inventors of electron microscopy) the Nobel Prize for Physics in 1987.

STM relies, as its name implies, on the effect known as quantum mechanical tunnelling. The electron distribution of a solid conductor overlaps the surface and extends (slightly) into free space. If two solids are brought so close that their electron distributions overlap then a current (the tunnelling current) can flow if a potential is applied between them. The magnitude of the tunnelling current depends very sensitively (exponentially) on the spacing between the solids; it may change by three orders of magnitude as the spacing changes by one atomic diameter. It is the strength of the tunnelling current which provides the signal in an STM.

The experimental layout is shown in Figure 7.2. A probe in the form of a very sharp point is scanned, mechanically, across the surface to be studied. As the height of the surface changes, because of surface steps or even the 'bumps' due to individual atoms, the tunnelling current changes. Its magnitude is sensed by the electronics and used to control a feedback system which alters the gap between the probe and the specimen until the pre-set tunnelling current is restored. This mechanism gives us an electron microscope with no 'primary beam', no 'secondary effect' and no lenses! The display effectively shows the

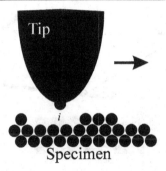

Figure 7.2 A very simplified diagram of a scanning tunnelling microscope (STM). As the tip is scanned across the specimen surface the tunnelling current (i) is modified. A feedback system is used to maintain the current, and hence the gap, at the same value. The tip thus follows the surface contours and its vertical displacement can be used to map the surface topography on a very fine scale.

vertical displacement of the tip as the specimen is scanned and is similar in many ways to a conventional SEM micrograph, showing a detailed image of the surface topography. It has already proved possible to image the positions of individual atoms on surfaces and thus to see directly what might be called the 'surface crystallography'.

It might seem amazing that the tip of the probe can be made sharp enough and that it can be moved sensitively enough through such small distances. In fact it has proved easier to achieve these feats than either the original designers, Binnig and Rohrer, or other early workers imagined. The probe does not have to be particularly sharp since only its most protruding atom will be brought to within tunnelling distance of the specimen. Movement of the probe in all three directions (x and y for scanning, z for control of the gap) is achieved by mounting it on piezoelectric crystals. Small voltage changes cause these crystals to expand or contract, thus moving the probe in a predictable fashion. Some precautions have to be taken to isolate the probe and specimen from vibration but these have proved not to be difficult. The STM has rapidly undergone a similar design evolution to most novel instruments; at first it seemed complex and was over-designed, then it was realized that the design could be simplified. STMs have been installed inside the chamber of a conventional SEM, inside semiconductor growth chambers and in other locations where they are at the heart of the experiment. No technique is totally idiot-proof but STM and its derivatives are now straightforward enough that any laboratory can use them.

Atomic force microscopy

Once the concept of the STM had been shown to work, the idea was quickly extended (and often simplified) to develop other techniques for revealing

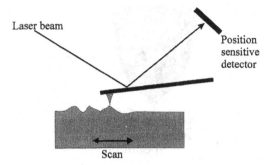

Figure 7.3 The elements of an atomic force microscope (AFM). Vertical deflections of the cantilever are detected as the reflected laser beam is deflected.

surface topography on a very fine scale. One of the techniques which has rapidly found widespread use is atomic force microscopy (AFM).

A fine tip, like that of an STM, is mounted on a cantilever beam. As this is brought up to the surface of the specimen it will be held at a constant distance by the interatomic forces between the atoms of the tip and those of the specimen surface. When the specimen is scanned, using piezoelectric crystals just as in the STM, the cantilever beam will rise and fall with the surface topography of the specimen. This vertical movement of the tip can be detected and measured very sensitively by reflecting a small laser beam from the cantilever, as shown schematically in Figure 7.3. The sensitivity of this technique is such that vertical deflections of significantly less than 0·1 nm can be detected, so surface steps only one atom high can be resolved quite easily. Lateral resolution on this scale is also possible, so the AFM genuinely has atomic resolution in all three dimensions. Figure 7.4 shows a typical image.

One amazing feature of AFM is that there is no special requirement for the specimen to be in a vacuum. In fact most AFM work is carried out in air, and good images can even be obtained under water or other liquids. The AFM therefore competes well with the SEM on cost and ease of use, and surpasses it in resolution. However all AFM images require computer processing and interpretation of these images is not as straightforward as with an SEM. Nor is its low magnification performance as good as that of an SEM, so both techniques will remain useful for many years.

A simple extension of the AFM is the magnetic force microscope (MFM) where the dominant force between tip and specimen results from their magnetic interaction. MFMs are now widely used to study magnetic materials.

Another clever development of the scanning probe idea also overcomes one of the apparent fundamental limits of microscopy using light. Near field optical microscopy (NFOM) uses a fine optical fibre tip to bring light to the scanned specimen. The light source is held so close to the specimen (as with STM and AFM) that the specimen is in the 'near field' and many of the usual rules of

1 μm

Figure 7.4 AFM image showing InAs islands on a GaAs substrate. (Lim Chee Han, University of Liverpool)

optics no longer apply. The physics of this is complicated, but essentially if both the diameter of the tip which is emitting the light and the distance between the tip and the specimen are smaller than the wavelength of the light, then microscopy can be performed using light at a resolution much smaller than the value of about half its wavelength implied by equation 1.4. This concept has been shown to work, but microscopes based on the NFOM principle are difficult to use and are still only to be found in a few research laboratories.

7.1.3 Field ion microscopy

The TEM is not yet able to image single point defects in crystalline specimens (although the STEM used in Z-contrast mode is almost there). The presence of vacancies and interstitial atoms has largely been deduced indirectly. The field ion microscope (FIM) permits some individual atoms and point defects to be imaged and thus extends the range of imaging techniques. In the FIM a high potential is applied between a very fine pointed specimen of tip radius r and a screen at a distance R (Figure 7.5(a)). In the intervening space a low pressure of inert gas is maintained: gas atoms are ionized in the very high field near the surface of the specimen. Each ion is then accelerated towards the screen, where it contributes a speck of light to the image. Ionization occurs preferentially

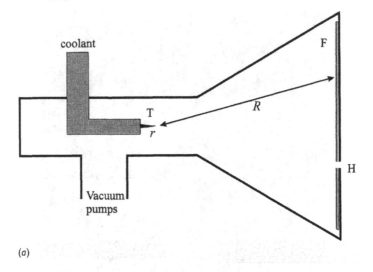

(a)

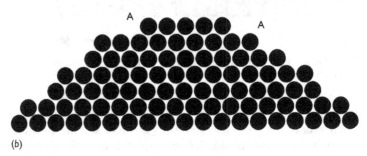

(b)

Figure 7.5 (a) The schematic layout of a field ion microscope. A high voltage is applied between the fluorescent screen (F) and the fine tip of the specimen (T). For atom probe analysis ions, which have been evaporated from the specimen tip, can pass through the hole (H) and enter a time-of-flight mass spectrometer. (b) Atom positions in a field ion microscope tip. Ionization of the imaging gas occurs above protruding atoms such as those marked A.

above those atoms which protrude from the specimen surface (e.g. at A in Figure 7.5(*b*)) and the image on the fluorescent screen thus consists of a pattern of bright spots corresponding to prominent individual atoms in the specimen. Two examples of FIM micrographs from crystalline specimens are shown in Figure 7.6. The magnification of an FIM image is simply R/r and hence in order to make atoms visible (i.e. to make their image bigger than 0.1 mm) r, the radius of the tip of the specimen, must be less than 1 µm. In fact the radius is likely to be significantly smaller than 1 µm; although images such as those shown in Figure 7.6 can be used to study single vacancies and grain boundary or dislocation structures in terms of atomic arrangements, only

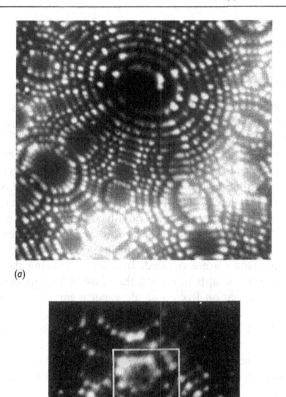

Figure 7.6 Field ion micrographs from (*a*) tungsten and (*b*) iridium. Image (*a*) shows the presence of a dislocation, since the spots lie on a spiral, not in concentric circles. Image (*b*) shows two vacant lattice sites as dark spots in the centre of the white box (J. A. Hudson, T. J. Godfrey and G. D. W. Smith)

a minute region of the specimen can be studied. The method is restricted to specimens with an electrically conducting or semiconducting matrix (so that the potential can be applied) and there are obvious difficulties in ensuring that the structures observed are representative of the bulk material.

Nevertheless the FIM offers one of the only simple ways of extending the resolution available in the TEM down to atomic dimensions. An extension of the technique (APFIM) also offers atom by atom analysis, and this is discussed in section 7.2.1.

7.1.4 Developments in electron microscopy

Many of the recent advances in electron microscopy have been stimulated by two developments, neither of which has actually required any great advance in the microscope itself or in the physics which underlies it. The most obvious development has been in the power of computing, which has enabled microscopists to perform many feats which, although possible previously, would have taken far too long a time to be worthwhile. An example is the modern ability to subtract the background from an X-ray or EELS map (section 6.3.2) before it is displayed: no new principle is involved but without a computer such image manipulation would be ridiculously time-consuming. The second development was the realization that during 'ordinary' electron microscopy many potentially useful signals are generated. Instead of ignoring these signals the modern microscopist attempts to exploit them, using a multi-signal approach, to extract more information from a single experiment. As a very elementary example, thirty years ago the generation of characteristic X-rays or backscattered electrons from a TEM specimen was totally ignored, although both processes were well understood. Nowadays such analytical techniques are commonplace. Ten years ago it took, perhaps, one hour to solve a backscattered or Kikuchi pattern. Now this is done in 0.1 s and this has effectively made a new analytical technique available.

Scanning transmission electron microscope (STEM)

The best known development in electron microscopy in recent years has been the STEM. This technique, as has been discussed in earlier chapters (sections 4.4 and 5.9.3), combines TEM and SEM by collecting a transmission image by the scanning method, and is ideally suited to a multi-signal approach. Since the resolution of a scanning technique is limited by the diameter of the probing beam, in order to achieve good resolution with a STEM a very fine electron probe is needed. It has proved possible, using the field emission electron gun (FEG, section 2.3), to generate electron probes which carry a current approaching 10^{-10} A in a beam of diameter only 0.2 nm $(2 \times 10^{-10}$ m$)$. This is sufficient to image fairly heavy single atoms in ideal non-crystalline specimens, and this achievement was responsible for the enthusiastic development of STEM instruments.

By invoking the principle of reciprocity (section 1.4) it is possible to show that, if the conditions of beam convergence and imaging aperture angle are equivalent, a STEM image should show identical contrast to a conventional TEM image. It was hoped in the early days of STEM that images could be obtained from thicker specimens than was possible in a TEM. This apparent

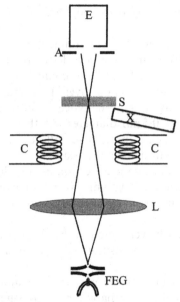

Figure 7.7 A schematic diagram of a scanning transmission microscope (STEM). The field emission gun (FEG) at the bottom provides a beam of electrons which is focused by the lens(es) L and scanned by the coils C before interacting with the specimen S. In the sample chamber around the specimen there may be an X-ray detector X, an energy loss spectrometer E and an annular dark field electron detector A in addition to the conventional transmitted electron detector.

advantage is illusory since images from a thick specimen show degraded resolution. If the desired resolution is specified then a STEM image of a crystalline specimen will be in no way superior to a CTEM image. For imaging purposes the TEM has therefore regained the ascendancy, since it offers a more useful combination of imaging and diffraction. The major current use of the so-called 'dedicated' STEM (sometimes called the FEGSTEM) is as a multi-signal analytical microscope. Details of the technique are given in Keyse *et al.* (1998).

Since the field emission gun requires a very high vacuum (about 10^{-11} mbar or 10^{-9} Pa), dedicated STEM microscopes are generally designed with an ultra high vacuum (UHV) specimen chamber. Since a STEM needs no objective or projector lenses after the specimen (Figure 7.7) there is usually a large space available for a variety of signal detectors. It is not unusual to find an energy loss spectrometer, an annular dark field detector and an X-ray detector mounted near the specimen. The annular dark field detector enables a signal to be collected from electrons scattered in any direction through more than a specified angle. This gives a much bigger signal than does conventional dark field imaging (section 4.2.2) and is particularly useful for amorphous specimens. The annular dark field image from a crystalline specimen would obviously be created from more than one diffracted beam and is therefore not simple to interpret.

Since the specimen is in an UHV environment it should not suffer contamination during examination and therefore long counting times can often be used. The ability of a dedicated STEM to count for long times on very small areas, using a high brightness beam, gives it the best spatial resolution for analysis in the EM family. Useful analytical information has been obtained from regions as small as 0·5 nm diameter. At this level the stability of the specimen is absolutely crucial. In order to collect a meaningful analysis from a region of diameter 0·5 nm neither the beam nor the specimen must drift by more than about 0·1 nm during the collection period. Since the signal is bound to be small from such a small volume we might like to collect for several hundred seconds. This imposes a constraint even more severe than that for imaging.

The STEM is particularly well suited to the 'Z-contrast' technique described in section 4.2.1, which gives compositional information at high resolution from an imaging technique.

Dedicated STEM instruments are now rarely manufactured, because they are expensive and the increasing capabilities of 'conventional' TEMs with field emission electron guns ('FEGTEMs') mean that they can deliver most of the requirements for high resolution analysis. However a new generation of 'super-STEM' instruments which incorporate aberration correction and deliver even better analytical resolution may be developed early in the new century.

Electron holography

Dennis Gabor invented holography in 1948 in order to improve the resolution of the EM. However electron holograms proved too difficult to produce and the first practical demonstration of the principle had to await the development of the laser. Recently further efforts have been made to harness the technique for its original purpose. In a manner exactly analogous to conventional laser holography, two coherent electron beams, one from the specimen and one acting as a reference beam, can be combined to form a hologram. The technique can only be made to work if the microscope is very stable and if the electron beam, from a field emission source, is very coherent. This places considerable constraints on the microscope and it is only with tremendous skill and patience that useful holograms can be produced. Several laboratories in the world have now succeeded in this task and interesting results are being produced, mainly on the distribution of magnetic and electric fields in materials.

Imaging techniques compared

It is useful to compare the various imaging techniques in terms of their ability to resolve detail laterally (i.e. in the plane of the specimen) and simultaneously

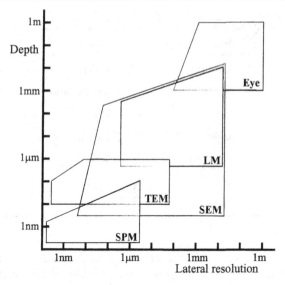

Figure 7.8 A map indicating the range of possible resolutions and the depth from which the image can be collected for the techniques discussed in the text. Notice that many techniques appear to have overlapping capabilities, particularly in the 10–100 nm resolution range. However, each technique does not necessarily image the same part of the specimen. For example SEM and TEM largely overlap but one is a surface technique while the other can image internal structure.

in the third direction (usually thought of as depth). We have done this in Figure 7.8 by plotting 'lateral resolution' against 'depth'. The fields shown for the techniques overlap considerably, as we might expect. Note that 'depth' is not always exactly equivalent to the depth of field defined in Chapter 1. For instance the limit to the depth information accessible to a TEM is not set by the notional depth of field but by the thickness of specimen which can be penetrated. In order to present the capabilities of such a wide range of instruments on a single diagram both scales are logarithmic, so each technique spans several orders of magnitude.

7.2 Complementary analysis techniques – alternative analysis systems

Very few microscopic analysis techniques are currently in widespread use. A few years ago it would have been possible to say that only the excitation of characteristic X-rays by high energy electrons had been substantially exploited. The statement is less true now but there still remains a huge potential for new alternative methods of analysis. This can be illustrated if we consider the nature of most physical analysis techniques – a primary probing beam (chosen from ten or more possibilities) is used to excite up to ten secondary effects locally

Table 7.1 Some of the possible imaging and analysis techniques.

Primary probe	Secondary effect	Variable	Technique
Electron	Electron	Energy	EELS
X-ray	X-ray	Temperature	XPS
Ion	Ion	Mass	SIMS
Light	Light	Intensity	SEM
Neutron	Neutron	Time	SAM
Sound	Sound	Angle	
Atom	Heat	Phase	

from the region of interest – the chosen secondary effects are then monitored as a function of one or more of seven or eight variables. At the crudest level we can combine these figures to arrive at the possibility of $10 \times 10 \times 7 = 700$ single-signal techniques, of which EM imaging, diffraction and X-ray analysis are only three. Of course there are many more possibilities for multi-signal techniques: more than a hundred of the single-signal techniques have been tried, but very few multi-signal techniques are yet in widespread use. Table 7.1 shows just a few of the possibilities. For most microscopic techniques we are interested in the variation of the detected variable as a function of position on the specimen, but this is not always the case. There is only space in this chapter to mention a few of the more popular techniques which currently complement electron microscope-based analysis.

7.2.1 Atom probe field ion microscope (APFIM)

One of the most sensitive analysis techniques yet invented has been developed from the field ion microscope. If a high potential or a short laser pulse is applied briefly to an FIM tip (i.e. the pointed specimen) a layer of atoms can be stripped from the surface. These will be ionized and will then accelerate towards the imaging screen (Figure 7.5(a)). In order to analyse a selected atom it is only necessary to orientate the specimen so that the image of the selected atom falls over a hole in the fluorescent screen, leading to a time-of-flight mass spectrometer. A sequence of single atoms can then be analysed by applying a series of high voltage pulses to the tip. Although the statistics of analysis in the atom probe are inevitably poor, and the technique is essentially destructive, the spatial resolution could scarcely be bettered. The APFIM technique has been used to great effect in materials science, as the example shown in Figure 7.9 illustrates.

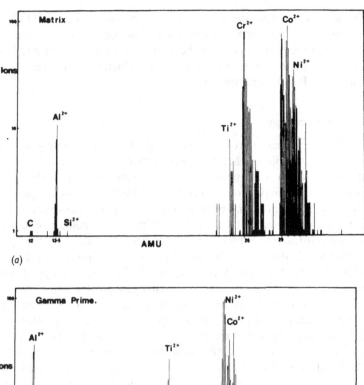

(a)

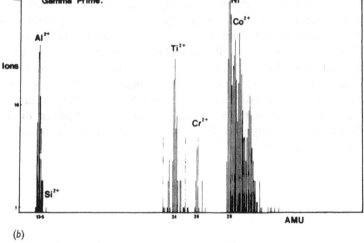

(b)

Figure 7.9 Atom probe spectra from a superalloy. Each spectrum shows the number of ions detected with each atomic mass (plotted as atomic mass units, AMU). Spectrum (a) shows that the matrix contains Ni, Co, Al and Ti; spectrum (b) was collected from a small gamma prime particle and it can be seen that it is significantly depleted in chromium. (G. D. W. Smith, University of Oxford)

During the 1990s the atom probe was developed into a truly three-dimensional analytical technique by the incorporation of position sensitive detectors which enable the detection and analysis of ions leaving any region of the tip, rather than just those at one location. With such a PoSAP (Position Sensitive

Atom Probe) and its associated computer processing facilities 3D elemental maps of small regions a few tens of nm across can be displayed. This is still a very small part of any specimen, but bear in mind that a region just 20 nm in each direction contains about a million atoms, which will have been analysed one by one by a PoSAP instrument.

7.2.2 STM spectroscopy

The scanning tunnelling microscope can be used not only simply to map the topography of a sample surface but also to investigate the electronic energy levels at any point on that surface. In the best circumstances, therefore, the nature and state of a single atom can be investigated. This type of spectroscopy is done by varying the potential between the STM tip and the specimen. Figure 7.10 shows, very schematically, how this affects the tunnelling processes and therefore the information which can be gleaned. The atom at the specimen surface will have a density of states similar to that shown in the diagram. Some of these states (those with energies below the Fermi energy, E_f) will be filled with electrons while others, with energies greater than E_f, will be empty. With no potential applied between tip and surface (Figure 7.10(a)), no tunnelling current will occur. If the specimen is biased positive (Figure 7.10(b)) then electrons can tunnel from the tip into empty states in the specimen. As the bias is varied the tunnelling current will change, telling us about the local density of unfilled states. If the specimen is biased negative (Figure 7.10(c)) then electrons can tunnel from filled states in the specimen to empty states in the tip. As we vary the negative bias we therefore learn about the filled electron states of the specimen. As this over-simplified discussion reveals, STM spectroscopy has the potential for extremely detailed and extremely localized analysis.

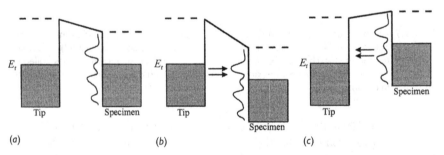

Figure 7.10 Tunnelling possibilities in an STM with (a) no bias, (b) positive bias and (c) negative bias.

7.2.3 SIMS

Ion-based surface analytical techniques are popular because of their sensitivity and their ability, in some cases, to reveal the variation of composition with depth beneath the original specimen surface (the depth profile). In particular, secondary ion mass spectrometry (SIMS) is extremely useful. In this technique a beam of primary ions, which can be focused down to a diameter of about 20 nm and can be scanned, is used to eject secondary ions from the specimen.

The mass (or strictly charge to mass ratio) of each secondary ion is determined by a mass spectrometer. The technique is clearly 'destructive' in that the layer of atoms being analysed is removed from the specimen. However, at low primary ion beam currents this occurs very slowly and the technique is known at 'static SIMS'. In favourable cases as little as 0·1% of a monolayer of material can be detected. If higher primary ion beam currents are used, material is removed more rapidly and depth profiling can be achieved since each successive layer of atoms can be analysed as it is 'peeled away'. This is known as 'dynamic SIMS'. Modern SIMS equipment is capable of a spatial resolution better than 1 μm and therefore composition maps can be displayed, which appear similar to the X-ray maps shown in Figures 6.10 and 6.11. This is often called 'imaging SIMS'. All SIMS techniques enjoy two great advantages. The first is element range: since a mass spectrometer is sensitive to all elements (strictly all isotopes) even the 'difficult' light elements from hydrogen to oxygen can be analysed and mapped. The second advantage is sensitivity: SIMS is usually able to detect concentrations in the ppm (parts per million) range and in good circumstances can be sensitive to parts per billion (ppb). Remember that 1 ppm is 0·0001%.

7.2.4 RBS: Ion beam analysis

Most of the techniques which have been developed for use with scanned electron beams can be carried out using ions or protons. Ions generally penetrate the specimen much less deeply than electrons of equivalent energy, so they are more surface-sensitive. High energy ions excite secondary electrons and characteristic X-rays as well as secondary ions, so ion-based analogues of the SEM and microprobe analyser have been developed. In general the peak-to-background ratio for secondary effects excited by ions is larger than that for electron excitation so the sensitivity (i.e. the minimum detection limit) of ion-based techniques is better. On the other hand the spatial resolution of ion probes is generally worse than that of electron probes. A further range of techniques based on the scattering of high energy ions is becoming more widely used. The prime example is Rutherford Backscattering (RBS), in which helium ions (alpha particles) of 2 or 3 MeV are directed at the specimen. A fraction of the primary ions interact strongly with the nucleus of a specimen atom and they are deflected through more than 90 degrees. The

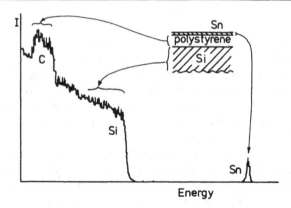

Figure 7.11 An RBS spectrum resulting from the scattering of a beam of 1·5 MeV helium ions from a silicon specimen on which has been deposited 200 nm of polystyrene and 5 nm of tin. The peak positions indicate the atomic number of the species, with heavier atoms giving smaller energy losses and appearing on the right of the spectrum. (R. S. Payne University of Surrey)

energy spectrum of these backscattered ions (e.g. Figure 7.11) can reveal a great deal about the nature of the specimen. Backscattered ions lose a well-defined amount of energy in their Rutherford collision, and also lose more by electronic interactions during their passage through the specimen. A spectrum such as Figure 7.11 therefore enables us to deduce not only the nature of the atoms in the specimen but also their depth distribution. Since the specimen is largely undamaged by MeV alpha particles this can be achieved non-destructively.

In favourable circumstances RBS can determine depth distributions to a resolution of 10 nm and can detect p.p.m. amounts of impurities. If the specimen is a single crystal, as semiconductors often are, then it can be set at such an orientation to the particle beam that the ions are 'channelled' along particular crystallographic directions. In this condition the RBS technique is even more sensitive to imperfections which locally modify the crystallinity, or to interstitial impurity atoms which are not on normal lattice sites.

7.2.5 Auger microscopy and spectroscopy

The emission of characteristic (Auger) electrons was described in section 2.8.3. The energies of the most useful Auger electrons are generally fairly low (a few keV) and they can be determined using an electron spectrometer of different design from the energy loss spectrometer described in section 6.6. The two most popular spectrometers are of the serial type and focus electrons of the selected energy electrostatically using either a cylindrical 'mirror' or a pair of hemispheres. If one of these spectrometers is mounted on a scanning electron microscope it should be possible to perform scanning Auger microscopy (SAM) and to create compositionally mapped images.

Despite the fact that the yield of Auger electrons can be quite high, especially from light elements (see section 2.8.3), most of them will be re-absorbed within a solid specimen and will not escape. High electron beam currents are therefore needed to generate a useful Auger signal and this means that the primary beam spot size must be quite large (see section 5.5.1). The resolution of Auger maps is therefore limited to about 50 nm. Nonetheless the SAM technique is a very powerful tool in the surface analyst's armoury. It would appear to be attractive to detect the Auger signal from the specimen in a TEM. However there are a number of practical problems. The space around the specimen is very restricted in a TEM and it is difficult to mount an Auger electron spectrometer in such a way that it can collect an appreciable fraction of the electrons emitted. The generated signal is, in any case, small since the majority of electrons in a TEM do not excite an inner-shell electron (see section 2.7.4). For these reasons only a very small number of laboratories in the world have yet succeeded in performing Auger analysis in a TEM.

7.2.6 Surface analysis by XPS

The surface analysis technique of scanning Auger microscopy has already been described in section 7.2.5. There are several other well established complementary techniques dedicated to the analysis of surfaces. Monochromatic X-rays or ultraviolet light are often used to excite photo-electrons whose energy carries information about the binding energy within their original atom, and thus about the nature of the atoms in the specimen. The characteristic kinetic energy of the photo-electron is given by the energy of the X-ray (or ultraviolet) photon, which is totally absorbed in the process, minus the characteristic binding energy of the electron which is emitted. These photo-electron spectroscopy techniques are called XPS and UPS or more loosely ESCA (Electron Spectroscopy for Chemical Analysis). Since photo-electrons can only escape from near the specimen surface the chemical information which they carry is specific to the top few atomic layers. Photo-electron energies are also sensitive to the state of bonding of the atom from which they come, so for instance metallic iron can be distinguished from iron in an oxide. XPS is well suited to determining composition and bonding state in thin layers near the specimen surface but it does not offer good spatial resolution because the exciting beam of X-rays covers a wide area and cannot readily be focused.

7.2.7 Raman spectroscopy

Raman scattering involves the inelastic scattering of light. Light scattered from a specimen shows either a shift to longer wavelengths (the Stokes shift, corresponding to energy loss) or a shift to shorter wavelengths (the anti-Stokes shift, corresponding to energy gain). The spectrum of reflected light, containing various Stokes and anti-Stokes lines, acts as a fingerprint for the chemical species on the specimen surface. Raman spectroscopy is thus a sensitive analytical technique, which can give information not just about elemental species but also about their bonding states. It is often carried out via a spectrometer attached to a light microscope, which is of course a well developed and convenient way of delivering light to and from the specimen. Increasingly Raman spectroscopy is being done with a confocal microscope (see section 7.1.1), making it quite easy both to identify the region of interest and to perform scanning to produce chemical maps.

7.2.8 Developments in EM

No totally new EM analysis techniques have been developed in the last ten years. However we can expect that both EDX and EELS techniques will continue to develop and that the Z-contrast STEM technique (see section 4.2.1) will become more widely used.

7.2.9 Analysis techniques compared

A graphical comparison of several analytical techniques is shown in Figure 7.12. This is a similar style of presentation to the imaging comparison in Figure 7.8 but instead of plotting image depth we now plot the depth (from the surface) from which the analytical information comes. The conclusion to be drawn is similar to that drawn from the earlier comparison. No one technique

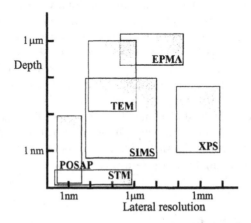

Figure 7.12 A map showing the spatial origin of the analytical information for many of the techniques discussed in the text. Notice that there may be significant extra differences between techniques in terms of their element range and sensitivity.

can solve all problems – many approaches are needed. That message has, we hope, come across clearly in the whole book.

7.3 Complementary diffraction techniques

There are many well established and widely used X-ray diffraction techniques which are complementary to the electron diffraction methods outlined here. Details of them are beyond the scope of this book. Neutron diffraction also retains a niche role for materials scientists.

7.3.1 RHEED and LEED

A discussion of electron diffraction would not be complete without mention of RHEED (Reflection High Energy Electron Diffraction) and LEED (Low Energy ED). Both these techniques are used to reduce the depth below the surface from which the diffraction information is collected. In RHEED this is achieved by directing the primary beam at a glancing angle of incidence to the specimen. Its penetration below the surface is therefore low, as Figure 7.13 shows.

The LEED approach is different in that the primary beam is usually incident perpendicular to the specimen surface but very low energies are used to reduce penetration. Extremely detailed surface structural information can be deduced

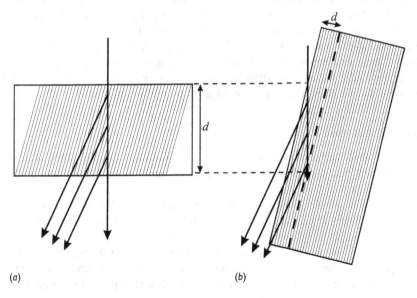

Figure 7.13 Electron diffraction from a crystalline specimen (*a*) in transmission and (*b*) in reflection. The effective thickness of material from which the diffraction information is collected, *d*, is much reduced by using a reflection technique.

from LEED patterns, although interpretation is not very easy. The highest resolution surface imaging techniques (REM, SREM, surface profiling and STM, see above and below) are now challenging LEED as the most convenient way to pinpoint atomic positions on surfaces.

7.3.2 Developments in EM diffraction

One area in which we can expect further development is the energy filtering of diffraction patterns in the electron microscope. We have discussed in section 6.6.4 the increasing use of energy filtering microscopes to produce images resulting from particular energy losses and thus particular elements. Energy filtered diffraction patterns are also very useful, because it is possible to remove inelastically scattered electrons (since these have by definition lost some energy). The resultant diffraction patterns contain only elastically scattered electrons and much fine detail, normally lost within the inelastic background, can be recovered.

7.4 Summary

We have shown how many variations of electron microscopy there are, and how many other techniques are available. We should never assume that any of these is the 'ultimate technique': existing techniques will continue to be refined and extended; new techniques will continue to be devised (such as, in our working lives, STM, AFM and confocal microscopy), and perhaps most significant of all we will improve our ability to extract (and interpret) two or more signals from our specimens. What all these advances have in common is that they will be made by people who understand the physical principles of imaging and analysis, and these are what we have tried to impart throughout this book. We hope that we have, at least partially, succeeded.

7.5 Questions

1   List three techniques which use serial image collection and three which use parallel image collection. Into which category would you put field ion microscopy?

2   If a 1 cm cylinder of piezoelectric material is to be used to move a specimen by up to 10 μm to provide the scanning for an STM, what is the maximum strain which needs to be induced in the material?

3   The PoSAP provides a very sensitive analysis in that individual atoms of a particular element can be counted. The detection of a single atom in 10^6 would appear to offer a sensitivity of 0·0001% (1 ppm). Why would you not

use the PoSAP to determine the carbon content of a steel or the iron content of an aluminium alloy to $\pm 0.0001\%$?

4 'Static' SIMS can be used to detect and analyse as little as a tenth of a monolayer of a solid surface. If the analysed area is 1 mm $\times$ 1 mm estimate the number of atoms which would be involved in such an analysis.

Further reading

SEM

Goldstein, J. I., Newbury, D. E., Echlin, P., Joy, D. C., Fiori, C. and Lifshin, E. (1981) *Scanning Electron Microscope and X-ray Analysis*. New York: Plenum Press.

Joy, D. C. (1984) 'Beam interactions, contrast and resolution in the SEM', *J. Microscopy* 136: 241.

Chescoe, D. and Goodhew, P. J. (1990) 'The operation of transmission and scanning microscopes', *Royal Microscopical Society Handbook 20*. Oxford: BIOS.

Richards, B. P. and Footner, P. K. (1992) 'The role of microscopy in semiconductor failure analysis', *Royal Microscopical Society Handbook 25*. Oxford: BIOS.

Reimer, L. (1998) *Scanning Electron Microscopy*. Berlin: Springer-Verlag.

TEM

Williams, D. B. and Carter, C. B. (1996) *Transmission Electron Microscopy: A Textbook for Materials Science*. New York: Plenum Press.

Chescoe, D. and Goodhew, P. J. (1990) 'The operation of transmission and scanning microscopes', *Royal Microscopical Society Handbook 20*. Oxford: BIOS.

Goodhew, P. J. (1985) 'Thin foil preparation for electron microscopy', in A. M. Glauert (ed.) *Practical Methods in Electron Microscopy*, Vol. 11, Amsterdam: Elsevier.

Keyse, R. J., Garratt-Reed, A. J., Goodhew, P. J. and Lorimer, G. W. (1998) *Introduction to Scanning Transmission Electron Microscopy*. Oxford: BIOS.

Reimer, L. (1995) *Energy-filtering Transmission Electron Microscopy*. Berlin: Springer-Verlag.

Tonomura, A. (1993) *Electron Holography*. Berlin: Springer-Verlag.

Maunsbach, A. Afzelius, B. and Afzelius, B. (1998) *Biomedical Electron Microscopy: Illustrated Methods and Interpretations*, London: Academic Press.

TEM and diffraction

Hammond, C. (1997) *The Basics of Crystallography and Diffraction*. Oxford: Oxford Science Publications.

Bristol EM Group (1985) *Convergent Beam Diffraction of Alloy Phases*. Bristol: Adam Hilger.

Edington, J. W. (1975) 'Electron diffraction', *Practical Electron Microscopy*, Vol. 2. London: Macmillan.

Hirsch, P. B., Howie, A., Nicholson, R. B., Pashley, D. W. and Whelan, M. J. (1965) *Electron Microscopy of Thin Crystals*. London: Butterworth.

Loretto, M. H. (1984) *Electron Beam Analysis of Materials*. London: Chapman & Hall.

Thomas, G. and Goringe, M. J. (1979) *Transmission Electron Microscopy of Materials*. Fairfax: Tech Books.

Analysis

Egerton, R. F. (1996) *Electron Energy Loss Spectroscopy in the Electron Microscope*, 2nd edn. New York: Plenum Press.

Hren, J. J., Goldstein, J. I. and Joy, D. C. (1979) *Introduction to Analytical Electron Microscopy*. New York: Plenum Press.

Reed, S. J. B. (1993) *Electron Microprobe Analysis*, 2nd edn. Cambridge: Cambridge University Press.

Scott, V. D. and Love, G. (1983) *Quantitative Electron-probe Microanalysis*. Chichester: Ellis Horwood.

Other microscopies

Amelinckx, S., van Dyck, D., van Landuyt, J. and van Tenderloo, G. (eds) (1997) *Handbook of Microscopy*. Weinheim: VCH.

Sheppard, C. J. R. and Shotton, D. M. (1997) 'Confocal laser scanning microscopy', *Royal Microscopical Society Handbook 38*. Oxford: BIOS.

Bradbury, S. and Bracegirdle, B. (1998) 'Introduction to light microscopy', *Royal Microscopical Society Handbook 42*. Oxford: BIOS.

Matter Software

All the software referred to in this book is to be found in the MATTER modules contained in *Materials Science on CD-ROM*, Liverpool University Press or Chapman & Hall (1998). For details of upgrades and later releases see www.liv.ac.uk/~matter/home.html or www.matter.org.uk

Answers

Chapter 1

1 Equation 1.1 gives 2 mm
2 At infinity
3 6
4 Equation 1.4 gives $r = 709$ nm
5 Equations 1.1 and 1.2 give 40 cm from lens, magnification $= 4$
6 $\tan \alpha = 0 \cdot 5/20$ then from figure 1.11 $h = d/\tan \alpha = 40 \, \mu$m. Depth of focus = 10 000 greater $= 40$ cm
7 2, 3, 4
8 Length e.g. mm
   For EM equation 1.9 with 100 kV electrons gives 953 nm, almost 1 μm.
   For light assuming $0 \cdot 5 \, \mu$m wavelength gives $0 \cdot 23 \, \mu$m.
9 Because after passing through the specimen some electrons must have lost energy, thereby changing their wavelength and contributing to a spread of wavelengths in the beams which pass through the objective and projector lenses

Chapter 2

1 Equation just above 2.5 gives $j = 5 \times 10^7 \, \mathrm{Am}^{-2}$ or $5 \times 10^3 \, \mathrm{Acm}^{-2}$. This is $3 \cdot 1 \times 10^8$ electrons nm^{-2}s^{-1}
2 W is close to $0 \cdot 5$ for $Z = 32$ (Germanium)
3 Equation 2.3 gives $5 \cdot 3 \times 10^{-31}$ kg (almost a 60% increase)
4 Auger energy should be smaller, because the departing electron has to overcome (i.e. lose) its binding energy
5 $P(0) = 0 \cdot 358$; $p(3) = 0 \cdot 093$ so 64·2% are not scattered while 9·3% are scattered three times.
6 99·4%
7 3.31×10^{-33} m
8 $1s^2 \, 2s^2 \, 2p^6 \, 3s^2 \, 3p^6 \, 3d^{10} \, 4s^1$

9 From equation 2.11 wavelength $= 0 \cdot 61$ nm; could be $P(K_\alpha)$, $Zr(L_\alpha)$ or possibly $Pt(M_\alpha)$

10 Advantages: vibrational stability, room at top for access to viewing chamber (image detector EELS, etc). Disadvantage: harder to access gun

Chapter 3

1

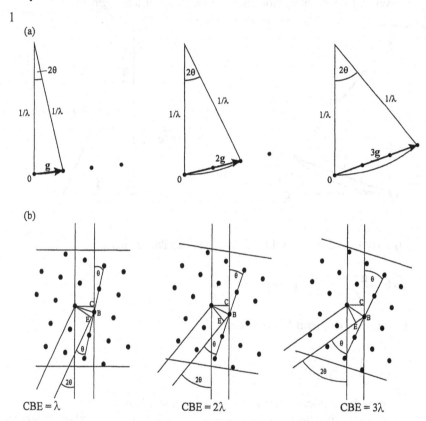

(a)

(b)

CBE $= \lambda$ CBE $= 2\lambda$ CBE $= 3\lambda$

2 First order, $9 \cdot 18$ mrad ($0 \cdot 526°$); second order $18 \cdot 35$ mrad ($1.052°$)

3 Spots become closer together; region of zero order Laue zone expands; HOLZ rings increase in diameter

4 Use equation 3.6
   Primitive cubic: 100; 110; 111; 200; 210
   Body centred cubic: 110; 200; 211; 220; 310

5 (a) $2\bar{2}0$; $\bar{2}20$; $20\bar{2}$; $\bar{2}02$; $02\bar{2}$; $0\bar{2}2$.
   (b) From equation 3.9 and Table 3.1, $a = 0 \cdot 4243$ nm

6 (a) By inserting a selected area aperture over the grain in the image, using parallel illumination

(b) By focusing the electron beam down to a small probe using C2 and placing the beam on the grain using the beam deflectors

7　An increase in diffuse intensity between spots; loss of detail in HOLZ lines in CBED patterns

8　From section 3.4.1, 0·22 mrad (0·0124°)

9　In selected area diffraction, the HOLZ rings are made up of spots, whilst in convergent beam diffraction they are made up of line segments.

10

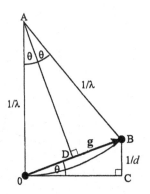 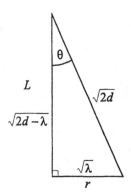

(a) From the figure, AOD and OBC are similar triangles.

$$\sin \theta = \frac{g/2}{1/\lambda} = \frac{1/d}{g}, \ \text{giving } \mathbf{g} = \sqrt{\frac{2}{\lambda d}}$$

Bragg's law can then be re-written

$$\sin \theta = \frac{\lambda g}{2} = \sqrt{\frac{\lambda}{2d}}$$

Squaring both sides of this equation gives $\lambda = 2d \sin^2 \theta$.

(b) Ignoring the effects of lens distortions, the length measured on the screen, r, is given by $L \tan \theta$, where L is the camera length.

Since $\sin \theta = \sqrt{\dfrac{\lambda}{2d}}$, by constructing the triangle above we find that

$$\tan \theta = \sqrt{\frac{\lambda}{2d - \lambda}}$$

and so the radius of the ring is

$$r = L\sqrt{\frac{\lambda}{2d - \lambda}}.$$

Chapter 4

1 Underfocus (smallest convergence angle), in focus, overfocus (largest convergence angle)

2 12.5 mrad

3 Decreasing the focal length of C1 increases the convergence angle and demagnification of the electron source, i.e. it decreases the spot size at the specimen. If C2 is not adjusted, it also increases the convergence at the specimen

4 From equation 3.4, the Bragg angle of the 002 beam is 10.28 mrad. For a parallel beam of electrons, this diffracted beam will be 10.28 µm away from the undeflected beam in the back focal plane, and would be blocked by a 20 µm aperture centred on the undeflected beam. For a convergent beam of electrons, the aperture would have to be smaller.

5 $100 \, \text{nm} \times 100 \, \text{nm} = 10^{-10} \, \text{cm}^2$.

6 (a) Just below the specimen entry port;
(b) about half way between the specimen and the screen;
(c) just below the electron gun.

7 In bright field imaging, the undeflected beam is used to form an image. The image will be bright in the absence of a specimen. In dark field imaging, a diffracted beam is used to form the image and the background is dark in the absence of a specimen (see also section 4.2.2 and Figure 4.12)

8 The contour will move, since a different part of the specimen will satisfy the Bragg condition for these planes

9 A dislocation is invisible if $\mathbf{g} \cdot \mathbf{b} = 0$, and a stacking fault is invisible if $\mathbf{g} \cdot \mathbf{R} = n$, n being any integer. This gives:
(a) dislocations visible, stacking fault invisible;
(b) dislocations invisible, stacking fault visible;
(c) both dislocations and stacking fault invisible

10 Mass thickness contrast, diffraction contrast, and phase contrast

Chapter 5

1 Reference to Figure 5.11 and equation 5.2 shows that the probe diameter (d) is reduced if the strength of the condenser lens is increased. In so doing, the angle α_0 is increased and thus the beam current as given by equation 5.3 is reduced. Increasing the size of aperture (A) would increase α_1 and therefore the beam current. However, this will increase the spherical aberration (equation 5.10) and this will ultimately limit the smallest spot size obtainable

2 The depth of field (h) at a particular magnification is determined by the working distance and objective aperture. From equation 5.8, it is found that an aperture no larger than 20 µm is required

3 The optimum operating conditions are when the beam diameter (d) is equal to the specimen pixel size (p) as given by equation 5.4 and therefore d should be

increased as the magnification (M) is reduced. If $d > p$ a blurred image results and if $d < p$ then the beam and hence the beam current are unnecessarily small and a noisy image may result

4   From equation 5.22 we find the backscattering coefficients (η_1 and η_2) to be 0·0343 and 0·0257. The natural contrast (C) is then given by equation 5.23 as 0·251

5   The beam current (I) is obtained from equation 5.20 as 2×10^{-11} A. d_{min} is obtained from equation 5.12 as 2·3 nm. Putting these values into equation 5.13 gives the minimum usable probe diameter (d) as 13·5 nm. For secondary electrons, this may be taken as the spatial resolution

Chapter 6

1   K for P
    L for Mo (K not efficiently generated)
    K for Cr (L not well detected)
    L or M for Os.

2   Be = 110/3·8 = 28
    Zr = 15770/3·8 = 4150
    Reasons include effect of noise/statistics on low number of counts i.e. smaller signal, broader peak, also spacing of peaks which are closer at low Z

3   (a) Ca
    (b) Ar
    Also Mg close to escape peak for Ar, Mn close to sum peak for Ar

4   $d > \lambda/2$ for Bragg's Law to work, so Fe > 0·095 nm
    Ti > 1·37 nm > only get such big spacings from artificial (soapy) crystals.

5   B = 61 nm so total diameter is 66 nm
    (a) smaller signal from particle at bottom than top (smaller proportion of electrons hit it)
    (b) lower spatial resolution of particles at bottom

6   Equation 6.4 gives 1·6%, 0·16% and 0·02%

7   $p(1)/p(0) = 0·1 = t/\lambda$, so $t = 12$ nm

Chapter 7

1   Series: from SEM, imaging SIMS, STM, AFM, confocal LM. Parallel from: TEM, conventional LM, GIF image filter, magnifying glass

2   0·1%

3   Inhomogeneity and sampling

4   Surface atoms about 0·2 nm square, so 0·1 ML contains about 2×10^{13} atoms. Comment: so statistics of counting should be good.

Index

Printed in the United States
by Baker & Taylor Publisher Services